The Economics of Ecosystems and Biodiversity in Business and Enterprise

This book is a product of the TEEB study (The Economics of Ecosystems and Biodiversity). It provides important evidence of growing corporate concern about biodiversity loss, and offers examples of how leading companies are taking action to conserve biodiversity and to restore ecosystems.

This book reviews indicators and drivers of biodiversity loss and ecosystem decline, and shows how these present both risks and opportunities to all businesses. It examines the changing preferences of consumers for nature-friendly products and services, and offers examples of how companies are responding. The book also describes recent initiatives to enable businesses to measure, value and report their impacts and dependencies on biodiversity and ecosystem services.

The authors review a range of practical tools to manage biodiversity risks in business, with examples of how companies are using these tools to reduce costs, protect their brands and deliver real business value. The book also explores the emergence of new business models that deliver biodiversity benefits and ecosystem services on a commercial basis, the policy enabling frameworks needed to stimulate investment and entrepreneurship to realize such opportunities, and the obstacles that must be overcome.

The book further examines how businesses can align their actions in relation to biodiversity and ecosystem services with other corporate responsibility initiatives, including community engagement and poverty reduction. Finally, the book concludes with a summary and recommendations for action.

The Economics of Ecosystems and Biodiversity in Business and Enterprise

Edited by Joshua Bishop

An Output of TEEB: The Economics of Ecosystems and Biodiversity

earthscan
London • New York

First published 2012
by Earthscan
2 Park Square, Milton Park, Abingdon, Oxon OX14 4RN

Simultaneously published in the USA and Canada by Earthscan
711 Third Avenue, New York, NY 10017

Earthscan is an imprint of the Taylor & Francis Group, an informa business

Disclaimer: The designations employed and the presentation of the material in this publication do not imply the expression of any opinion whatsoever on the part of the United Nations Environment Programme concerning the legal status of any country, territory, city or area or of its authorities, or concerning delimitation of its frontiers or boundaries. Moreover, the views expressed do not necessarily represent the decision or the stated policy of the United Nations Environment Programme, nor does citing of trade names or commercial processes constitute endorsement.

The official full citation for this volume is as follows: TEEB (2012) The Economics of Ecosystems and Biodiversity in Business and Enterprise. Edited by Joshua Bishop. Earthscan, London and New York.

British Library Cataloguing in Publication Data
A catalogue record for this book is available from the British Library

Library of Congress Cataloging in Publication Data
The economics of ecosystems and biodiversity in business and enterprise / edited by Joshua Bishop.
p. cm.
Includes bibliographical references and index.
1. Environmental economics. 2. Social responsibility of business. 3. Biodiversity. I. Bishop, Joshua.
HC79.E5E2783 2012
658.4'083--dc23
2011027826

ISBN: 978-1-84971-251-4 (hbk)
ISBN: 978-0-203-14170-0 (ebk)

Typeset in Sabon
by Domex e-Data Pvt. Ltd. (India)

Printed and bound in Great Britain by the MPG Books Group

Contents

CLARENCE HOUSE

I have, for many years, been increasingly concerned that we are damaging and over-consuming the planet's natural capital, on which we all depend for our prosperity and, indeed, survival. Experts tell me that even with over a billion people with no access to drinkable water and living on less than a dollar a day, we are consuming, every year, fifty per cent more of the planet's natural resources than it can renew. And, remarkably, we are depleting and destroying this wonderful asset with little idea of the economic consequences. This is, in no small part, due to the economic invisibility of Nature and the fact that the value of the planet's ecosystems and biodiversity has not been taken into account, fully and consistently, in our economic, accounting and decision-making systems.

We need to remember that the ultimate source of all economic capital is natural capital and that the world economy is a subsidiary of Nature, and not the other way round.

This is why I am so full of admiration for the tireless efforts made by Pavan Sukhdev and his team to produce the T.E.E.B. study and to publish this book. I hope that it will be an important step in raising awareness of the serious risks faced by business in not responding to biodiversity loss and ecosystem degradation – but also of the opportunities to be gained through an enlightened and far-sighted response. The fact that the Earth's irreplaceable natural capital has not, for the most part, been valued and priced is a critical gap in our ability to understand the true underlying structure of our economic systems. Indeed this is synonymous with having not adequately priced the costs of climate change, which is perhaps the world's greatest ever market failure, as Lord Stern pointed out in his seminal 2006 report. I am, as a result, delighted that this vital work is being taken forward by the T.E.E.B. For Business Coalition, supported by my Accounting for Sustainability Project.

I am sure that readers of this book will be inspired to see the opportunities for business that can be secured if the urgently needed mechanisms to value Nature's ecosystems and biodiversity are developed in time. These valuation mechanisms are vital if we are to meet the challenges of the twenty-first century.

List of Figures, Tables and Boxes

Figures

Tables

Boxes

List of Contributors

Overall editor

Joshua Bishop (International Union for Conservation of Nature)
While writing and editing this book, Joshua Bishop was Chief Economist at the International Union for Conservation of Nature (IUCN), based in Switzerland. He has co-authored several Earthscan books on the links between economics, the environment and sustainable development. His recent work has focused on translating the concepts and methods of environmental economics into the language and practice of business.

Editors

Nicolas Bertrand (United Nations Environment Programme)
William Evison (PricewaterhouseCoopers)
Sean Gilbert (Global Reporting Initiative)
Linda Hwang (Business for Social Responsibility)
Mikkel Kallesoe (World Business Council for Sustainable Development)
Cornis van der Lugt (United Nations Environment Programme)
Francis Vorhies (Earthmind)

Contributing authors

Roger Adams (Association of Chartered Certified Accountants)
Robert Barrington (Transparency International UK)
Wim Bartels (KPMG Sustainability)
Nicolas Bertrand (United Nations Environment Programme)
Joshua Bishop (International Union for Conservation of Nature)
Gérard Bos (Holcim Ltd)
Luke Brander (Institute for Environmental Studies)
Giulia Carbone (International Union for Conservation of Nature)
Ilana Cohen (Earthmind)
Michael Curran (Swiss Federal Institute of Technology, Zürich)
Andreas Drews (Deutsche Gesellschaft für Internationale Zusammenarbeit)
Emma Duncan (independent)
Jas Ellis (PricewaterhouseCoopers)
Eduardo Escobedo (United Nations Conference on Trade and Development)
William Evison (PricewaterhouseCoopers)
John Finisdore (World Resources Institute)
Naoya Furuta (International Union for Conservation of Nature)
Kathleen Gardiner (Suncor Energy Inc.)
Sean Gilbert (Global Reporting Initiative)
Marcus Gilleard (Earthwatch Europe)
Julie Gorte (Pax World)
Annelisa Grigg (Global Balance)

Scott Harrison (BC Hydro)
Stefanie Hellweg (Swiss Federal Institute of Technology, Zürich)
Joël Houdet (Orée)
Linda Hwang (BSR)
Cornelia Iliescu (United Nations Environment Programme)
Suhel al-Janabi (Deutsche Gesellschaft für Internationale Zusammenarbeit)
Lorena Jaramillo (United Nations Conference on Trade and Development)
Mikkel Kallesoe (World Business Council for Sustainable Development)
Chris Knight (PricewaterhouseCoopers)
Thomas Koellner (Bayreuth University)
Matthew Lynch (World Business Council for Sustainable Development)
Alistair McVittie (Scottish Agricultural College)
Ivo Mulder (United Nations Environment Programme Finance Initiative)
Tim Ogier (PricewaterhouseCoopers)
Nathalie Olsen (International Union for Conservation of Nature)
Jérôme Payet (SETEMIP-Environnement)
Jeff Peters (Syngenta AG)
Conrad Savy (Conservation International)
Christoph Schröter-Schlaack (Helmholtz Centre for Environmental Research – UFZ)
Bambi Semroc (Conservation International)
Brooks Shaffer (Earthmind)
Fulai Sheng (United Nations Environment Programme)
James Spurgeon (Environmental Resources Management)
Franziska Staubli (Swiss Import Promotion Programme)
Jim Stephenson (PricewaterhouseCoopers)
Peter Sutherland (GHD)
Rashila Tong (Holcim Ltd)
Mark Trevitt (Trucost plc)
Alexandra Vakrou (European Commission)
Cornis van der Lugt (United Nations Environment Programme)
Francis Vorhies (Earthmind)
Christopher Webb (PricewaterhouseCoopers)
Olivia White (PricewaterhouseCoopers)

Other contributors and reviewers

Naoki Adachi, Margaret Adey, Mubariq Ahmad, Juan-Marco Alvarez, Maia Ambegaokar, Annika Andersson, Stuart Anstee, Simon Anthony, Gigi Arino, Geanne van Arkel, Kit Armstrong, Paul Armsworth, Andrea Athanas, Bruce Aylward, JiSu Bang, Edward Barbier, Lara Barbier, Monica Barcellos, Steve Bartell, Ricardo Bayon, Desiree Beeren, Uwe Beständig, Sheila Bonini, Maria Ana Borges, Roberto Bossi, Sue Both, David Brand, David Bresch, John Brown, Tim Buchanan, Jürg Busenhart, Jim Cannon, Nathaniel Carroll, Peter Carter, Catherine Cassagne, Sagarika Chatterjee, Garrette Clark, Claus Conzelmann, Polly Courtice, Toby Croucher, Valerie David, Mark Day, Andrea Debbane, Andrew Deutsch, Laksmi Dhewanthi, Ian Dickie, Elaine Dorward-King, Sophie Dunkerley, Derek Eaton, Evelyn Ebert, Steinar Eldoy, Edgar Endrukaitis, Jan Fehse, Frauke Fischer, Anne-Marie Fleury, Hans Friederich, Kaori Fujita, Peter Gardiner, Franz Gatzweiler, Sandra Geisler, James Gifford, Sean Gilbert, Juan Gonzalez-Valero, Sara Goulartt, James Griffiths, Dolf de Groot,

Moustapha Kamal Gueye, Matt Hale, Jun Hangai, Derek de la Harpe, Celia Harvey, Anida Haryatmo, Tetsu Hattori, Hazel Henderson, Wiebke Herding, Frank Hicks, Kii Hiyashi, Paul Hohnen, Gemma Holmes, Takashi Hongo, Ard Hordijk, David Huberman, Salman Hussein, Mira Inbar, Rufus Isaacs, Tilman Jaeger, Ian Jameson, Lorena Jaramillo, Tsukasa Kanai, Sachin Kapila, Stefanie Kaufmann, Nijma Khan, Paola Kistler, Carla Kleinjohann, Adam Klimkowski, Alan Knight, Paula Knight, Ayoko Kohno, Ryo Kohsaka, Andreas Kontoleon, Nicolas Kosoy, Eszter Kovács, Pushpam Kumar, Georgina Langdale, Alistair Langer, Rik Kutsch Lojenga, Paula Loveday-Smith, Nadine McCormick, Jennifer McLin, Jeff McNeely, Becca Madsen, Andy Mangan, Tony Manwaring, Joseph Mariathasan, Kiyoshi Matsuda, S. Matsuura, Richard Mattison, Eva Mayerhof, Aditi Mehta, Susanne Menzel, Andrew Mitchell, Narina Mnatsakanian, Cristina Montenegro, Jennifer Morris, Herman Mulder, Katrina Mullan, Nobuo Nakanishi, Carsten Nessöver, Aude Neuville, Tim Nevard, Gijsbert Nollen, Paulo A. L. D. Nunes, Maria-Julia Oliva, Yoko Otaki, Michael Oxman, Olivia Palin, Sandra Paulsen, Ashim Paun, Helena Pavese, Paola Pedroni, Chris Perceval, Danièle Perrot-Maître, Gergana Petrova, Sander van der Ploeg, Wendy Proctor, Deric Quaile, Mohammad Rafiq, Kurt Ramin, Irene Rankin, Rob Regoort, Alison Reinert, Dave Richards, Steven Ripley, Ruth Romer, Per Sandberg, Stefan Schaltegger, Oliver Schelske, Dorothea Seebode, Andrew Seidl, Delia Shannon, Ravi Sharma, Jennifer Shaw, Paul Sheldon, Benjamin Simmons, Anthony Simon, Josselyn Simpson, Paul Simpson, Daniel Skambracks, Karin Skantze, Tim Smit, Kerstin Sobania, Laura Somerville, Naoko Souma, Richard Spencer, Nina Springer, Dale Squires, Franziska Staubli, Susan Steinhagen, Vladimir Stenek, Harve Stoeck, Simon Stuart, Virpi Stucki, Wataru Suzuki, Tomomi Takada, Kazuaki Takahashi, Anislene Tavares, Bouwe Taverne, Patrick ten Brink, Brian Thomson, Celine Tilly, Lloyd Timberlake, Mathieu Tolian, Jo Treweek, Juan Carlos Vasquez, James Vause, Marcos Vaz, Olivier Vilaca, Donn Waage, Sissel Waage, Tom Watson, Jacques Weber, Mark Weick, Jeffrey Wielgus, Bernd Wilke, Elizabeth Willetts, Jon Williams, Britt Willskytt, Heidi Wittmer, Kaori Yasuda, Giuseppe Zaccagnini

Acknowledgements

This book is the product of many hands, hearts and minds working together over some three years on a study of The Economics of Ecosystems and Biodiversity (TEEB). It is virtually impossible to recognize every individual who helped make TEEB a success. At the beginning and end of each chapter we list the contributors to that chapter, including the editors and lead writers, contributing authors, reviewers and others who provided contacts, documents or other helpful inputs. Here we acknowledge those who contributed more generally to plan, prepare and communicate the 'TEEB in Business' report, which formed the basis of this book.

We acknowledge first the vision and encouragement provided by the leaders of several organizations, who saw the strategic importance and potential impact of TEEB and gave it their full support from its earliest conception through delivery of the final Synthesis report in late 2010. These include notably Sigmar Gabriel and Jochen Flaschbart (German Ministry for Environment, Nature Conservation and Nuclear Safety), Stavros Dimas and Ladislav Miko (European Commission), Julia Marton-Lefèvre (IUCN), Achim Steiner (UNEP), and several other members of the TEEB Advisory Board, not to mention the delegates of the G8+5 countries who provided the original political mandate for TEEB, in the form of the 2007 Potsdam Initiative (www.bmu.de/files/english/pdf/application/pdf/potsdam_initiative_en.pdf).

Among their many contributions, these leaders made an inspired choice by appointing an extraordinary Study Leader: Pavan Sukhdev provided consistent, demanding and encouraging leadership to a dispersed and sometimes fractious team, and he has been a tireless ambassador and eloquent spokesman for TEEB throughout the study. Other key institutional and practical supporters at critical stages of the study include: Jock Martin and Ronan Uhel (EEA), William Jackson, Jeffrey McNeely and Juan-Marco Alvarez (IUCN), Malcolm Preston, Chris Knight and Jon Williams (PwC), Ahmed Djoghlaf and Ravi Sharma (SCBD), Hussein Abaza and Benjamin Simmons (UNEP), James Griffiths and Bjorn Stigson (WBCSD).

Like other TEEB 'deliverables', the study for business was coordinated by a 'core group', which included the lead editors and writers of this book, as well as other technical advisers. Members of this group are acknowledged elsewhere in this book, in the lists of contributors to each chapter. Here we recognize other individuals who provided more general support for the work, including initial project development, stakeholder outreach and administrative assistance. In no particular order, they are: Sissel Waage (BSR), Andreas Kontoleon and Katrina Mullan (Cambridge University), Francis Vorhies and Brooks Shaffer (Earthmind), Tim Hardwick and Ashley Irons (Earthscan), Guy Duke, Aude Neuville, Alexandra Vakrou and Stephen White (European Commission), Paolo Nunes (FEEM), Patrick ten Brink (IEEP), Catherine Cassagne (IFC), Haripriya Gundimeda (Indian Institute of Technology, Bombay), Andrea Athanas, Sue Both, Evelyn Ebert, Naoya Furuta, Maria Hasler, Padma Lal, Nadine McCormick, Alex Moiseev, Brian Thomson, Elizabeth Willetts and Sebastian Winkler (IUCN), Pushpam Kumar (Liverpool University), Sheila Bonini and Josselyn Simpson (McKinsey), Kii Hayashi (Nagoya University), Jun Hangai (Nippon Keidanren), Gemma Holmes and Paula Knight (PwC), Naoki Adachi (Response Ability), Tony Manwaring and Aditi Mehta (Tomorrow's Company), Augustin Berghofer, Florian Eppink, Johannes Forster, Melanie Heyde, Carsten Nesshover, Christoph Schröter-Schlaack and Heidi Wittmer (UFZ), Lara Barbier, Garette Clark, Georgina Langdale, Désiree Leon, Paula Loveday-Smith, Rahila Mughal, Sarah Odera, Fatma Pandey, Mark Schauer and Susan Steinhagen (UNEP), Rudolf de Groot and Sander van der Ploeg (Wageningen University), Jacques Weber and Barbara Bendandi, among many others.

Finally, for helping to ensure that the TEEB results do not just sit on bookshelves but lead to wider discussion, deeper reflection and more effective influence on business decisions, we acknowledge the efforts of the many individuals and organizations that have tried to get the message out and promote follow-up initiatives building on TEEB in Business, including Helena Pavese (Conservation International Brazil), Peter Carter (European Investment Bank), Mikael Salo (Miljoaktuellt), Jussi Soramaki (Finnish Ministry of Environment), Richard Spencer (Institute of Chartered Accountants of England and Wales), Giulia Carbone, Hastings Chikoko, Kurt Ramin, Andrew Seidl and Jaeger Tilman (IUCN), Willem Ferwerda and Rob Regoort (IUCN Netherlands Committee), Chris Mahon (IUCN UK National Committee), Sandra Paulsen and Karin Skantze (Swedish Environmental Protection Agency), John Brown (Ten Alps), James Vause (UK DEFRA), Jon Hutton (UNEP-WCMC), Beatrice Otto (WBCSD), Martina Gmur, Jason Shellaby and Tom Watson (WEF), James Gifford and Narina Mnatsakanian (UNPRI), and Herman Mulder, among many others.

To all those not listed above who gave no less of their valuable time and energy to ensure the success of TEEB and the completion of this book, we can only offer our most humble apologies and sincere thanks.

Preface

Pavan Sukhdev, Study Leader, TEEB
Joshua Bishop, Coordinator, TEEB in Business

The idea of placing monetary values on biological diversity ('biodiversity') and of using economic incentives to encourage the delivery of ecosystem services is controversial. Some people question whether it is meaningful or indeed ethical to translate life on Earth into the language of economics and business. How can we evaluate ecosystem costs and benefits when there are no realistic substitutes for the living fabric of this planet? Do we know enough about how nature functions to support economic analysis or to design effective responses? Even more controversially, can or should we entrust the management of nature to business and markets? Even if the ethical and scientific questions about biodiversity valuation can be resolved, what could possibly persuade governments, businesses and society in general to 'internalize' such values, especially given that the costs incurred are often in the form of public goods and services, with no markets, prices or property rights?

Such questions are valid, but the fact remains that non-economic arguments for, and approaches to, biodiversity conservation have shown themselves to be insufficient. The 'status quo' is unethical, unsustainable and in need of urgent corrective action, while most current conservation practice is simply ineffective in the face of rapid economic change.

The premise of TEEB, a global study of the economics of ecosystems and biodiversity, may be compared with recent studies of the economics of climate change (e.g. Stern 2006). The starting point for TEEB is the fact that the fundamental drivers of environmental decline are to be found in the realm of economics. Moreover, like recent studies of the economics of climate change, TEEB highlights the key role of economic institutions, policies and actors – notably markets, economic incentives and the business community – in delivering effective solutions to biodiversity loss.

One of the main reasons for the widespread losses of biodiversity and degradation of ecosystems that we observe today is the economic invisibility of nature. While we measure economic gains in market assets and income, the decline in environmental quality remains largely unrecorded on the balance sheets of business entities, governments and even households. To entrust our future to a decision making framework in which trade-offs are made without recognizing losses on one side of the ledger cannot be a sound strategy for human progress.

For example, the decision to convert natural ecosystems to agriculture or residential development cannot be considered economically efficient if it is based (implicitly or explicitly) on the idea that there is no price (and hence no value) for the ecosystem services lost due to land-use change. And yet this is exactly what we observe in most cases of forest clearance to create more cattle pasture or to grow more palm oil.

Externalizing environmental costs while internalizing profits – destroying *public wealth* while creating *private wealth* – cannot be the basis of sustainable growth for business or of prosperity for society as a whole. A report by the UN-backed Principles for Responsible Investment (based on research by Trucost) indicates that the 3,000 largest publicly listed companies in the world imposed almost US$2.15 trillion of environmental damages to society in 2008, mostly in the form of greenhouse gas emissions, unsustainable use of fresh water, pollution, etc. This was equivalent to 7 per cent of the combined revenues of these same corporations in the same year.

While some sectors bear more responsibility than others, all businesses are implicated and vulnerable to the effects of biodiversity loss and ecosystem decline. As with climate change, we can anticipate major changes ahead, whether by design – through deliberate efforts to address the problem – or by default, as society is forced to adapt to the impacts of environmental change.

A proactive approach to biodiversity loss requires much closer attention to the status of and trends in natural capital. It also entails increased investment in ecosystem conservation and restoration, expanding the reach of markets through payments for ecosystem services and economic incentives that encourage resource-efficient production and consumption, and a shift in the burden of taxation from what we *make* (i.e. profits and jobs from producing goods and services) to what we *take* from nature (i.e. natural resource degradation and pollution), among other changes.

The dominant economic model today appears to promote *more* rather than *better* consumption, the creation of *private* over *public* wealth, and building *man-made* capital instead of *natural* capital. This 'triple-whammy' of self-reinforcing biases leads to an economy in which we extract resources without fear of natural limits, consume without awareness of environmental consequences, and produce without responsibility for external costs. This is the exact antithesis of a 'green economy', which would improve human well-being and reduce inequalities while also reducing environmental risks and ecological scarcities.

The challenge, as always, is managing change, including its unintended consequences. The TEEB approach favours pragmatism over perfectionism, planned changes over reaction, common sense and attention to social equity over 'free market fundamentalism'. The instruments proposed by TEEB include reforms to existing economic policies and environmental regulations, as well as the introduction of new financing mechanisms and ecosystem markets, where appropriate. A common theme in all TEEB reports is the need for greater recognition and more explicit valuation of nature's benefits to society. Economic valuation is needed to communicate the value of nature to decision makers in their own language, which is dominated by economic concepts and paradigms.

This volume focuses on decision makers in the world of business. It is accompanied by a separate volume on the ecological and economic foundations of TEEB (TEEB Foundations 2010), which synthesizes the 'state of the art' in the economic valuation of nature, as well as two other volumes, for national and international policy makers (TEEB in Policy Making 2011), and for local and regional policy makers and resource managers (TEEB in Local Policy 2011). In addition, TEEB has developed a dedicated website and other resources for individuals (http://bankofnaturalcapital.com/).

This volume argues that there are both serious risks to business, as well as significant opportunities, related to biodiversity loss and ecosystem degradation. There is a need for all businesses to quantify and value their impacts on biodiversity and ecosystems, in order to avoid or mitigate risks and maximize positive opportunities. Evaluations of any kind are a powerful 'feedback mechanism' for a society that has distanced itself from the biosphere, upon which its very health and survival depends. Economic valuations, in particular, communicate the value of ecosystems and biodiversity and their largely unpriced flows of public goods and services in the language of the world's dominant economic and political model. Mainstreaming this thinking and bringing it to the attention of policy makers, administrators, businesses and citizens, is the central purpose of TEEB. This volume on TEEB in Business and Enterprise is a contribution towards that objective.

References

Stern, N. (2006) *The Economics of Climate Change*, Cambridge University Press, Cambridge.

TEEB Foundations (2010) *The Economics of Ecosystems and Biodiversity: Ecological and Economic Foundations* (ed. P. Kumar), Earthscan, London.

TEEB in Local Policy (2011) *The Economics of Ecosystems and Biodiversity in Local and Regional Policy and Management* (ed. H. Wittmer and H. Gundimeda), Earthscan, London.

TEEB in Policy Making (2011) *The Economics of Ecosystems and Biodiversity in National and International Policy Making* (ed. P. ten Brink), Earthscan, London.

List of Acronyms and Abbreviations

ABS	access and benefit sharing
ACC	Aquaculture Certification Council
AFOLU	Agriculture, Forestry and Other Land Use
ARIES	ARtificial Intelligence for Ecosystem Services
BAP	biodiversity action plan
BBOP	Business and Biodiversity Offsets Program
BDP	biodiversity damage potential
BES	biodiversity and ecosystem services
BMS	Biodiversity Management System
CBD	Convention on Biological Diversity
CCBS	Climate, Community and Biodiversity Standard
CDM	Clean Development Mechanism
CDP	Carbon Disclosure Project
CDSB	Climate Disclosure Standards Board
CEO	chief executive officer
CESR	Corporate Ecosystem Services Review
CI	Conservation International
CITES	Convention on International Trade in Endangered Species of Wild Fauna and Flora
CNFCM	Center for Natural Forest Conservation Management
CSR	corporate social responsibility
DEFRA	Department for Environment, Food and Rural Affairs (UK)
DFID	Department for International Development (UK)
EBI	Energy and Biodiversity Initiative
EEA	European Environment Agency
EIA	environmental impact assessment
EJF	Environmental Justice Foundation
ELD	Environmental Liability Directive (EU)
EMA	Environmental Management Accounting
ESB	Ecosystem Services Benchmark
ESDP	ecosystem services damage potential
ESHIA	environmental, social and health impact assessment
FAO	Food and Agriculture Organization of the United Nations
FEEM	Fondazione Eni Enrico Mattei
FFI	Fauna & Flora International
FLO	Fairtrade Labelling Organizations International
FMCG	fast-moving consumer goods
FPP	Forest, Paper & Packaging
FSC	Forest Stewardship Council
GAA	Global Aquaculture Alliance
GACP	Good Agricultural and Collection Practices
GDI	Green Development Initiative
GHG	greenhouse gas
GMO	genetically modified organism

GRI	Global Reporting Initiative
GWT	Global Water Tool
HCV	High Conservation Value
IBAT	Integrated Biodiversity Assessment Tool
ICMM	International Council on Mining and Metals
IEA	International Energy Agency
IEEP	Institute for European Environmental Policy
IFAW	International Fund for Animal Welfare
IFC	International Finance Corporation
IFOAM	International Federation of Organic Agriculture Movements
ILCD	International Reference Life Cycle Data System
IOAS	International Organic Accreditation Service
IPCC	Intergovernmental Panel on Climate Change
IPIECA	International Petroleum Industry Environmental Conservation Association
ISEAL	International Social and Environmental Accreditation and Labelling Alliance
ISO	International Organization for Standardization
ISRP	Independent Scientific Review Panel
IUCN	International Union for Conservation of Nature
JBIB	Japan Business Initiative for Conservation and Sustainable Use of Biodiversity
JISL	Jain Irrigation Systems Ltd
LCA	life cycle assessment
LCI	life cycle inventory
LCIA	life cycle impact assessment
LCM	life cycle management
LOHAS	lifestyles of health and sustainability
MA	Millennium Ecosystem Assessment
MAC	Marine Aquarium Council
MIS	micro-irrigation systems
MSC	Marine Stewardship Council
NBSAP	National Biodiversity Strategy and Action Plan
NFCP	Natural Forest Conservation Program
NGO	non-governmental organization
NNL	no net loss
NPI	net positive impact
NVI	Natural Value Initiative
OECD	Organisation for Economic Co-operation and Development
PDCA	plan–do–check–act
PEFC	Programme for the Endorsement of Forest Certification
PES	payments for ecosystem services
PRI	Principles for Responsible Investment
PwC	PricewaterhouseCoopers
RA	Rainforest Alliance
REDD+	Reducing Emissions from Deforestation and Forest Degradation (including the conservation of forest carbon stocks, sustainable management of forests and enhancement of forest carbon stocks)
RSB	Roundtable on Sustainable Biofuels
RSPO	Roundtable on Sustainable Palm Oil
SAI	Sustainable Agriculture Initiative

SCBD	Secretariat of the Convention on Biological Diversity
SIA	social impact assessment
SMART	specific, measurable, achievable, relevant and time-bound
SME	small and medium-sized enterprise
SRI	socially responsible investing
STPR	Social Time Preference Rate
TEEB	The Economics of Ecosystems and Biodiversity
TIES	The International Ecotourism Society
TNC	The Nature Conservancy
UEBT	Union for Ethical BioTrade
UFZ	Helmholtz Centre for Environmental Research
UN	United Nations
UNCTAD	United Nations Conference on Trade and Development
UNDP	United Nations Development Programme
UNEP	United Nations Environment Programme
UNEP FI	United Nations Environment Programme Finance Initiative
UNFCCC	United Nations Framework Convention on Climate Change
VCS	Voluntary Carbon Standard
WACC	weighted average cost of capital
WBCSD	World Business Council for Sustainable Development
WCMC	World Conservation Monitoring Centre
WDBA	World Database of Protected Areas
WEF	World Economic Forum
WFTO	World Fair Trade Organization
WRI	World Resources Institute
WSSD	World Summit on Sustainable Development
WTP	willingness to pay
WTTC	World Travel and Tourism Council
XBRL	eXtensible Business Reporting Language

Chapter 1

Introduction to Biodiversity and Ecosystems for Business

Editors
Joshua Bishop (IUCN), William Evison (PricewaterhouseCoopers)

Contributing author
Olivia White (PricewaterhouseCoopers)

Contents

Key messages

The world is changing in ways that affect the value of biodiversity and ecosystem services to business. The value of biodiversity and ecosystem services (BES) is a function of population growth and urbanization, economic growth and ecosystem decline, changing politics and environmental policy, and developments in information and technology.

Biodiversity loss and ecosystem decline cannot be considered in isolation from other trends. The continuing loss of biodiversity and associated decline in ecosystem services is driven by growing and shifting markets, resource exploitation and climate change, among other factors. Equally, the loss of BES contributes to many of these other trends, implying the need for an integrated business response.

Business risks and opportunities associated with biodiversity and ecosystem services are growing. Given the ongoing decline of BES and the interaction between biodiversity loss, decline in ecosystem services and other major trends, business can expect both the associated risks and opportunities to increase over time.

There will be increasing pressure on (and more restricted access to) natural resources. Growing market demand for natural resources combined with increasing public concerns about environmental quality point towards increasing competition and more restricted access to natural resources on both land and sea.

Consumers increasingly consider biodiversity and ecosystems in their purchasing decisions. Consumer understanding and expectations of how products and companies relate to BES are becoming more sophisticated. Consumer-facing businesses, in particular, but also their suppliers, may need to re-examine how they manage BES and how their actions are communicated to customers.

Business is beginning to notice the threat posed by biodiversity loss. 27 per cent of global CEOs surveyed by PricewaterhouseCoopers (PwC) during the second half of 2009 expressed concern about the impacts of biodiversity loss on their business growth prospects. Interestingly, 53 per cent of CEOs in Latin America and 45 per cent in Africa expressed concern about biodiversity loss, compared with just 11 per cent in central and eastern Europe.

1.1 Background to this book

We live in a world transformed by business. Business has prospered by providing products and services to people everywhere, and business plays a key role in economic development. For nature, however, the price of development and business success has been very high.

Most people in business know about climate change and accept the need to reduce greenhouse gas emissions to levels that are consistent with a stable climate. Business leaders are also becoming more aware of the risks of biodiversity loss and the need to respect ecological limits generally (MA 2005).

The economic value of nature is changing, reflecting changes in people's preferences, demography, markets, technology and the environment itself. Companies are responding, but much more work is required to develop and scale up competitive business models that can conserve biodiversity and deliver ecosystem services while also meeting people's needs for better products and services.

The loss of biodiversity and valuable ecosystem services is increasingly well documented (see Chapter 2) and increasingly recognized as creating risks to business (Athanas et al. 2006). Business risk may be related to the direct impacts of a company's operations on biodiversity, or to the dependence of a business on ecosystem services as inputs to production. In other cases, the business risks associated with biodiversity loss may be indirect, operating through supply chains or through market decisions on investment, production, distribution and marketing (see Chapter 3). Companies around the world are finding ways to identify, avoid and mitigate their BES risks, using a range of new tools developed by, with and for business (see Chapter 4).

At the same time, biodiversity and ecosystem services are also the basis of new business opportunities (see Chapter 5). This is most obvious in the case of companies selling goods and services that are directly associated with biodiversity and ecosystems, such as nature-based tourism. But as with BES risk, there are less direct links between commerce and conservation that offer further opportunities. As a result, more and more investors and entrepreneurs are setting up funds and firms dedicated to building biodiversity business (Bishop et al. 2008). At the same time, some companies are discovering that integrating biodiversity and ecosystem services in their management systems can also help achieve wider corporate social responsibility goals (see Chapter 6).

The starting point for this analysis is the well-known fact that markets will not ensure efficient use of resources for which prices are lacking (TEEB Foundations 2010).[1] Because many of the benefits of BES are not reflected in the market prices of goods and services, often due to missing or poorly enforced property rights, these benefits tend to be neglected or undervalued in both public and private decision making. This leads to actions that result in biodiversity and ecosystem loss, which in turn may affect human well-being adversely. This book reviews the state of the art in measuring and managing biodiversity and ecosystem risks in business, capitalizing on new biodiversity business opportunities, and integrating business, biodiversity and development.

1.2 Approach, structure and contents

This section offers a preview of the remainder of this book. First, however, we define some key terms, identify our major assumptions, describe the methods used to compile this book, and list the main objectives and questions that this book seeks to address. We also identify the potential audience for this book and suggest where different readers will find material of interest to them. In the following section, the chapter turns to recent evidence of how business leaders and consumers think about biodiversity and ecosystems, and how this relates to other major trends affecting business.

1.2.1 Definitions

Throughout this book we use the terms biodiversity, ecosystems and ecosystem services, frequently abbreviated as 'BES'. These terms are defined as follows:

Biodiversity is short-hand for 'biological diversity'. We follow the Convention on Biological Diversity (CBD), which defines biodiversity as: 'The variability among living organisms from all sources including, inter alia, terrestrial, marine and other aquatic ecosystems and the ecological complexes of which they are part; this includes diversity within species, between species and of ecosystems' (Article 2).

According to the CBD, *ecosystems* are thus one component of biological diversity. This is consistent with definitions subsequently adopted by the Millennium Ecosystem Assessment (MA 2005), which identifies an ecosystem as 'a dynamic complex of plant, animal, and microorganism communities and the non-living environment interacting as a functional unit'. The main contribution of the MA is the elaboration of the concept of *ecosystem services*, defined simply as the benefits people receive from ecosystems (for further discussion see Chapter 2).

An important feature of ecosystem services is that they are culturally determined and therefore dynamic. As noted by TEEB, ecosystem services are 'conceptualizations ("labels") of the "useful things" ecosystems "do" for people, directly *and* indirectly ... whereby it should be realized that properties of ecological systems that people regard as "useful" may change over time even if the ecological system itself remains in a relatively constant state' (TEEB Foundations 2010: 18).

1.2.2 Assumptions

Turning from definitions to assumptions and from ecology to economics, this book adopts an explicitly economic perspective on the links between business, biodiversity, ecosystems and ecosystem services. This implies a focus on the value of natural resources to people, rather than on any 'intrinsic' value that may be ascribed to natural resources in their own right (e.g. a 'right to exist'). Of course, we recognize that many of the values people derive from BES are intangible, including recreational, cultural and 'existence' values, and that these intangible values may be significant. They are also measurable.

An economic approach further implies the acceptance of marginal trade-offs between BES benefits and other things that people value. While trade-offs may be constrained by lack of adequate substitutes for certain natural resources or ecosystem services, the fact remains that people do weigh the benefits of nature conservation against other things they value in life. In principle, if all values to people are fully reflected in such trade-offs, and subject to some other standard economic assumptions, we can be confident that the resulting use of resources will be economically efficient.

In practice, of course, the economic ideal of perfectly competitive markets, complete and instantaneous information, zero transaction costs, perfect substitution, complete property rights, etc. is never fulfilled. Nevertheless, we argue that more explicit consideration of BES costs and benefits in economic decision making generally results in better (if not optimal) outcomes. Economic valuation may never be perfectly accurate, especially where non-market values are at stake, but it is hard to think of decisions that are not improved by information about economic values, alongside other considerations.

Some other important assumptions behind this book should also be acknowledged:

- We assume continued economic growth and further integration of market-based democracies worldwide, alongside increased public awareness and concern about environmental change, and increased government regulatory capacity and constraints on the use of natural resources. While we acknowledge the existence of non-market models of economic organization, as well as non-democratic forms of government, we see no reason to doubt the continued growth of private enterprise, within increasingly subtle frameworks of economic policy, overseen by democratic governments and guided by increasingly well-informed citizens.
- We also recognize the growing economic and political power of several 'emerging' economies, and of companies headquartered in these countries. One notable feature of these emerging economies and companies is their apparent lack of explicit attention to environmental issues generally, and BES specifically, relative to more 'established' industrial economies and companies. While we have sought out developing country examples to illustrate our arguments throughout this book, we must acknowledge that the weight of documented experience (or claims) is concentrated in the developed world.
- Adopting an economic approach implies that incentives matter. In other words, property rights and prices influence human behaviour and the use of natural resources. The current failure of market incentives and public policy in most countries to reflect the full value of biodiversity and ecosystem services is one of the main reasons for continued loss of biodiversity and under-investment in natural capital. By the same token, effective action to conserve biodiversity and secure ecosystem services often requires creating or strengthening economic incentives for the conservation and sustainable use of biological resources.
- One corollary of this assumption is that purely charitable approaches to nature conservation, based on appeals to moral, ethical or religious values, are unlikely in market-dominated economies to mobilize significant private investment in biodiversity conservation. Although charity can make a real difference and should always be encouraged, any attempt to promote widespread, sustained and substantial private investment in nature conservation requires more forceful arguments, based on commercial logic and shareholder value.
- While charity may not be sufficient, the principles of free choice and voluntary action should be cherished and are essential features of an economic approach to nature conservation. Wherever possible, private firms and consumers should be allowed and encouraged to make mutually satisfactory environmental 'deals' on a voluntary basis, supported by legally binding contracts. Where such voluntary arrangements are not efficient, due to the presence of 'externalities' or other market distortions, governments can sometimes help by creating an enabling framework of incentives to encourage producers and consumers to

'internalize' environmental values in their transactions. What governments must avoid, however, are simplistic rules and regulations that ignore real differences in private costs and preferences, 'lock in' outmoded technologies or production practices, or otherwise undermine the constructive potential of business innovation.

- This last point relates to yet another key assumption: continued technological progress, stimulated in part by the increasing scarcity of natural resources and ecosystem services. Having said that, it is unlikely that technological innovation can compensate entirely and in every case for the loss of biodiversity and ecosystem decline. This implies that there are situations in which some type or level of environmental damage may be deemed unacceptable, irrespective of opportunity costs or the extent and quality of compensation provided. In short, marginal economic analysis does not apply to non-marginal events.
- Finally, we argue here that conservation and commerce can (and indeed must) work hand in hand if biodiversity loss and ecosystem decline are to be slowed and ultimately halted. While business is often responsible for environmental damage, efforts to make business part of the solution to biodiversity loss are likely to involve more business involvement in nature conservation and environmental management. This point, of course, is not easily proved. It is hoped that this book will go some way to showing, through multiple examples and practical inspiration, how increased environmental rights for business can be combined with increased responsibilities, such that commercial success is more closely aligned with conserving nature.

1.2.3 Methodology

The assumptions outlined above have guided our approach to compiling arguments and evidence for this book. In general, we have sought out examples that show how integrating BES in decision making can deliver real and tangible value to business, as well as positive environmental outcomes. Wherever possible, we have favoured case studies that provide financial and/or economic data on BES values.

Unfortunately, business records and reporting on BES are sketchy, anecdotal and inconsistent, making it difficult to paint a complete picture of biodiversity in business today. We rely heavily on case studies from those few companies that are willing and able to provide information about their policies and actions in relation to BES, along with a handful of independent assessments. Almost by definition, the case studies highlighted here are not from typical companies.

While our findings are therefore preliminary and incomplete, it is hoped that this book will stimulate more systematic and comprehensive study of business awareness, strategy and action in relation to BES. Such research is urgently needed to identify the most efficient means of stimulating business investment in biodiversity conservation, ecosystem restoration and sustainable use of natural resources.

1.2.4 Objectives and key questions

Given the limitations outlined above, what objectives did we set out to achieve and what questions do we seek to answer? In general terms, this book aims to present the best available evidence of the case for incorporating BES in business, including both risks and opportunities. In order to make this case and provide practical guidance to readers, the book compiles and summarizes multiple examples of how real businesses are using specific tools, techniques and initiatives to manage their relationship with BES and prepare for the future.

More specifically, the book addresses the following questions. How is the business and biodiversity context changing? How do business leaders perceive the risks of biodiversity loss? Will new technologies and emerging markets, as well as changing public policies and consumer preferences, alter the way that business values biological resources? (See Section 1.3.)

What is happening to biodiversity, what are the direct and underlying drivers of environmental change, and how does this affect business? What are the impacts and dependencies of different industry sectors on biodiversity and ecosystems? How do these impacts and dependencies create risks and opportunities for business? (See Chapter 2.)

How can business measure and report its impacts and dependence on biodiversity and ecosystems? Where does BES fit into corporate governance and management information? How can environmental information systems at site, product and group level be expanded to accommodate BES information? What is the experience of BES reporting in business and how can it be strengthened? (See Chapter 3.)

What are the risks of biodiversity and ecosystem loss to business and how can they best be managed? What tools are available to identify, assess and mitigate BES risks, and what value do they offer to business? What other methods and approaches can help companies reduce BES risk? (See Chapter 4.)

What are the main business opportunities related to biodiversity and ecosystems and how can they best be realized? How can BES be a value proposition today for existing industries? How can business make the best of emerging markets for biodiversity and ecosystem services? What tools and policies are available to support markets for biodiversity and ecosystem services? (See Chapter 5.)

How can business integrate action on BES with their commitments to development? What are the trade-offs and potential synergies between BES, socio-economic development and poverty reduction? What are the main barriers to integrating BES and poverty reduction, and what role can business play in helping to minimize trade-offs and maximize positive synergies? (See Chapter 6.)

Who needs to act and how in order to improve business–biodiversity relations? What guidance is available to business on biodiversity and ecosystems? What is the experience of voluntary action by business on BES and what lessons may be learned from other corporate responsibility initiatives? What are the main information gaps and other constraints on business action in favour of BES? (See Chapter 7.)

1.2.5 Audience

This book argues that biodiversity and ecosystems are valuable to all businesses, in every sector and country. Hence, the target audience for this book includes publicly traded companies and industry associations, state-owned enterprises and financial services, small and medium-sized enterprises, emerging companies from developing economies, business schools and others working at the interface between business and nature. The book considers a range of industry sectors, including agriculture, food and beverages, extractive industries, manufacturing, infrastructure and services.

Although detailed analysis of individual business sectors is beyond the scope of this study, preliminary assessments of BES impacts and dependencies, risks and opportunities are provided for a range of sectors (see especially Chapters 2 and 5). Moreover, in all chapters we try to include a selection of different sectors in the choice of case studies.

General readers interested in obtaining an overview of the status and trends in biodiversity, ecosystems and ecosystem services, focusing on links with business, should look at Chapter 2.

Those responsible for business environmental information systems and seeking to integrate BES data in corporate planning, accounting and reporting will find detailed discussion of current trends and tools in Chapter 3.

Project and product managers concerned about identifying and reducing BES risks should turn to Chapter 4 for practical guidance and examples.

Business planners, investors and entrepreneurs, as well as government regulators and development bankers, may find inspiration in Chapter 5, with its focus on BES as the basis of cost-savings, potential new products and the promise of new markets for biodiversity and ecosystem services.

Chapter 6 will interest those researching or responsible for corporate social responsibility in general, who may wonder how to integrate BES in business commitments to social development and poverty reduction.

Finally, Chapter 7 should be most relevant to those seeking an overview and comparative assessment of corporate social and environmental responsibility initiatives.

1.3 Biodiversity and ecosystems in a changing world

This book focuses on the links between business, biodiversity and ecosystem services. Of course, businesses are influenced by a range of social and economic factors, many of which also have implications for biodiversity and ecosystem services. Any attempt to improve the relationship between business, biodiversity and ecosystems must therefore take account of these wider factors and the linkages between them. This section examines several major trends that affect business, focusing on the linkages between these trends, BES risk and BES opportunity. First, however, we look at evidence of the current awareness and response of business leaders to biodiversity loss.

1.3.1 Perceptions of business leaders on biodiversity and ecosystems

A survey of 1,200 CEOs from around the world provides insight into current perceptions of the risk of biodiversity loss to business (PricewaterhouseCoopers 2010). When asked to rate their levels of concern about a range of threats to business growth prospects, 27 per cent of CEOs were either 'extremely' or 'somewhat' concerned about 'biodiversity loss'. Given the current economic context, it is striking that biodiversity loss is such a concern for some businesses. There is some interesting regional variation, with 53 per cent of CEOs in Latin America and 45 per cent in Africa expressing concern that biodiversity loss will adversely affect business growth prospects, compared with just 11 per cent in Central and Eastern Europe (Figure 1.1).

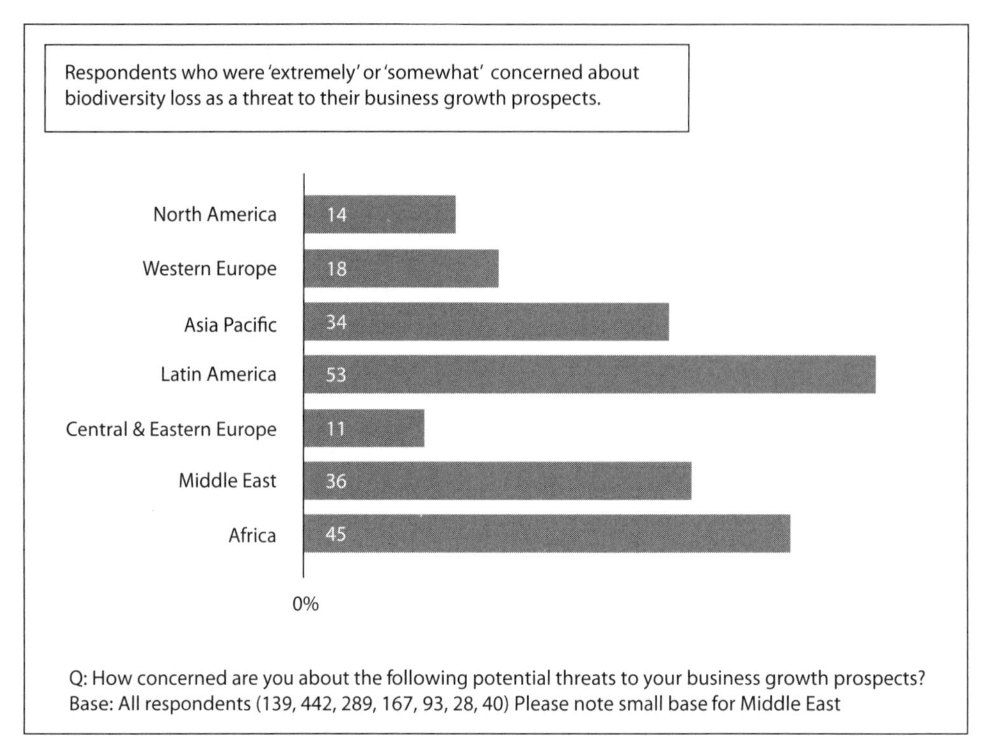

Figure 1.1 *Views of global CEOs on the threat to business growth from biodiversity loss*

Source: PricewaterhouseCoopers 13th Annual Global CEO Survey 2010

Nevertheless, at a global level CEO concern about biodiversity loss does not appear to be as high as for some other risks. For example, in the same survey, 65 per cent of CEOs expressed concern about a protracted global recession, 60 per cent about over-regulation, 54 per cent about energy costs, while 35 per cent had concerns about climate change.

These findings are supported by another 2010 survey of 1,500 executives (McKinsey & Company 2010); while nearly two-thirds of respondents to this survey reported that biodiversity was at least 'somewhat important' to their companies, on a list of 12 issues relating to the environment and sustainability, biodiversity ranked tenth, behind climate change, but also behind pollution and human rights.

The apparent lack of focus on biodiversity loss by business leaders may be partly due to lack of understanding of the potential implications for business. It may also reflect the fact that the effects of biodiversity loss and declining ecosystem services are not, in most cases, dramatic one-off events but rather a gradual trend, and so less visible to business leaders. In addition, as outlined below, the loss of BES may be overlain by other, more immediate trends and risks that are more visible to business leaders.

A striking finding of McKinsey's survey was that 59 per cent of respondents saw biodiversity as more of an opportunity than a risk for their companies. By comparison, just 29 per cent saw the issue of climate change as more of an opportunity than a threat in a similar survey conducted in late 2007 (McKinsey & Company 2007).

On the subject of regulation, PricewaterhouseCooper's survey revealed more scepticism than optimism among CEOs about the effectiveness of government action to protect biodiversity and ecosystems (Figure 1.2). What is less clear is the extent to which business leaders would like to see more government action, including regulatory reforms, to address biodiversity loss.

A third and more focused survey, of Japanese companies, carried out in early 2010, provides further insight on the level of business awareness and action on biodiversity in a major industrialized economy. This survey targeted 493 companies, of which 147 responded. The survey was designed as a follow-up to the 2009 'Declaration of Biodiversity' by a major Japanese business association and thus could be expected to reveal relatively high levels of awareness (Nippon Keidanren 2010). It is therefore perhaps not surprising that 50 per cent of respondents reported that they had already integrated 'biodiversity' into their company environmental policy, with a further 57 per cent of those who had not done so suggesting that they would in the future.

Among the same group of Japanese companies, 15 per cent confirmed that they have developed internal guidelines on biodiversity, with a further 42 per cent indicating that guidelines were either under development or planned. This survey demonstrates the impact that a business-led initiative, such as Nippon Keidanren's 'Declaration of Biodiversity', can have on corporate perceptions. Of course, recognition in company policy is only the first step towards effective management of biodiversity and ecosystem services. Because the initiative by Nippon Keidanren and its members is relatively new, clear evidence of improved BES outcomes is not yet available.

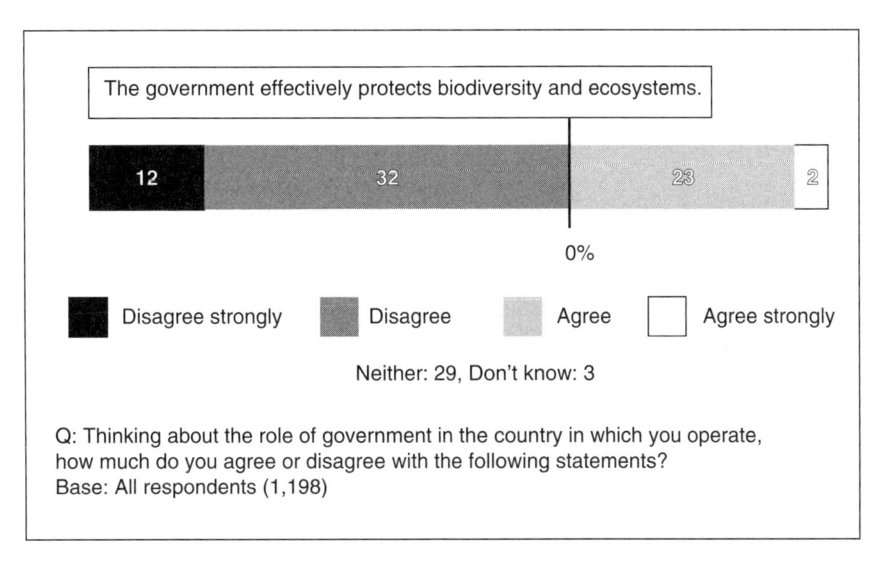

Figure 1.2 *Views of global CEOs on government protection of biodiversity and ecosystems*

Source: PricewaterhouseCoopers 13th Annual Global CEO Survey 2010

1.3.2 The emerging business environment: What trends matter?

Surveys of business leaders suggest limited awareness of the potential risks posed by biodiversity loss, although some companies in some countries are more concerned and have begun to respond. This and the following section explore some of the external factors – and the linkages between them – that may lead to more business awareness and action on biodiversity and ecosystem services in the coming years.

Visions of the future are inevitably uncertain. Nevertheless, several organizations have developed projections or scenarios across a range of themes and time periods, including climate change (Intergovernmental Panel on Climate Change – IPCC), energy (International Energy Agency – IEA), demographics (United Nations – UN), the environment and human well-being (United Nations Environment Programme – UNEP), food and water security (Food and Agriculture Organization of the United Nations – FAO), ecosystem health (Millennium Ecosystem Assessment – MA) and many other issues. All these efforts to explore the future have an element of truth to them. At the same time, past experience suggests that such forecasts are almost always inaccurate, due to our inability to anticipate significant changes, be they social, political, technological or environmental.

Such forecasts may be most useful as reminders of the major risks and opportunities that can affect business in the future, to which the appropriate response is not hard planning and irreversible commitments but rather investments in organizational resilience and adaptability.

One of the most comprehensive recent explorations of the future was a collaborative study led by the World Business Council for Sustainable Development, under the banner 'Vision 2050' (WBCSD 2010). Vision 2050 framed its analysis around the key factors or conditions that any plan for a more sustainable future needs to address, including:

- population growth and urbanization;
- economic growth and ecosystem decline;
- politics and environmental policy;
- information and technology.

These are briefly examined below, focusing on their relevance to biodiversity and ecosystems.

Population growth

According to the UN, the world population is expected to grow from 6.7 billion today to 9.2 billion by 2050. Of this increase, 98 per cent will occur in the developing world. The populations of developed countries are stabilizing and aging, a pattern that will eventually apply worldwide (the proportion of people aged over 60 has been rising steadily, passing from 8 per cent in 1950 to 11 per cent in 2007, and is expected to reach 22 per cent in 2050).

Population growth is expected to lead to increased demand for goods and services, and more pressure on natural resources. It is less clear what impact the aging of society will have on nature. However, the shifting balance of population implies that public perceptions of nature and the value of ecosystems may increasingly reflect the historical traditions and norms of developing rather than currently developed countries. These cannot be generalized but will probably reflect social attitudes to and human experience of nature in each region.

Urbanization

The world's urban population is expected to double by 2050, when about two-thirds of humanity will reside in cities. Urbanization suggests a more distant or indirect relation between people and nature, and thus perhaps greater emphasis on recreational, amenity and existence values of ecosystems and species, compared with more productive or utilitarian concerns. Urbanization also implies increased spatial concentration of some environmental impacts (e.g. residential and industrial land use, waste disposal, water pollution), as well as greater scope (and need) to establish payments for ecosystem services and other transfer mechanisms to capture and convey the willingness to pay of urban residents for resource stewardship by the remaining rural population.

Economic growth and ecosystem decline

Average income and consumption levels are generally increasing, mainly in the developing world. Continued reliance on carbon-based energy and accelerating use of natural resources will increase pressure on ecosystem services, threatening future supplies of food, fresh water, fibre and fish. According to Vision 2050, more than half the world's population is projected to live under conditions of severe water stress by 2025, while a larger proportion of the world's water use will be for irrigation. Simply meeting the demand for food for 9 billion people will require an increase in average crop yields of 2 per cent a year or more above recent levels.

WBCSD argues that economic growth must be 'decoupled from ecosystem destruction and material consumption, and re-coupled to sustainable economic development and meeting changing needs' (WBCSD 2010). The challenge is to ensure that such 'decoupling' does not simply imply the delocalization of adverse environmental impacts to distant production sites, but rather real improvements in the efficiency of energy and materials use. For instance, how can rising demand for animal protein be met without turning the world's remaining forests into pasture and feed crops? How can demand for mobility be satisfied without turning landscapes into motorways and car parks?

Politics

Current demographic and economic trends suggest that the developing economies will increasingly be on the front lines of efforts to achieve a sustainable future. According to WBCSD, the key challenge in the transition to sustainability is improving the quality of governance. As described in Vision 2050, governance systems should respect the principle of subsidiarity (i.e. decentralizing and making decisions at the most appropriate local level), but they must also 'pool sovereignty' where necessary to address international challenges such as trade, infectious disease, climate change, water resource management, high seas fisheries and other transboundary issues (WBCSD 2010).

According to WBCSD, future governance systems also need to be better at guiding markets to internalize environmental externalities, ensure transparency and inclusiveness, create a 'level playing field', and enable business to develop and deploy sustainable solutions. An outstanding question is whether the expected shift in economic and political power towards the larger emerging economies (i.e. the so-called BRICs) will result in new attitudes and approaches to environmental management, and ultimately help or hinder efforts to reach international cooperative agreements on managing the global commons.

Environmental policy

Reducing the environmental impacts of economic activity will entail changes in regulations, markets, consumer preferences, the pricing of inputs, and the measurement of profit and loss – all of which affect business. In the more 'sustainable future' envisaged by WBCSD, 'prices reflect all externalities: costs and benefits' (WBCSD 2010). This is seen as necessary to ensure that energy and resources are used efficiently and harmful emissions are reduced. For example, Vision 2050 proposes a 50 per cent reduction in greenhouse gas emissions by 2050 relative to 2005 levels, stimulated in part by public policy reforms that put a price on carbon (WBCSD 2010).

Such market-based approaches are increasingly being applied to other ecosystem services (besides climate regulation), implying that business can expect to pay more in future for its access to and impacts on a wide range of natural resources. At the same time, the adoption of market-based approaches to environmental management may imply greater business opportunities, based on the conservation of biodiversity and the provision or restoration of ecosystem services. Estimates developed by PricewaterhouseCoopers for Vision 2050 of 'sustainability related global business opportunities in natural resources (including energy, forestry, agriculture and food, water and metals)' suggest a potential market in the range of US$2–6 trillion by 2050 (at constant 2008 prices), about half of which is composed of 'additional investments in the energy sector related to reducing carbon emissions' (WBCSD 2010). While these estimates may be questioned, it seems likely that business will play an increasingly important role in the sustainable management of natural resources and the environment.

Information and technology

One of the biggest unknowns in attempts to predict the future is the pace and impact of technological change. To cite just one example, over 4 billion mobile phone handsets are now in use worldwide, three-quarters of them in the developing world. According to the World Bank, an extra 10 phones per 100 people in a typical developing country boosts GDP growth by almost one full percentage point, making a significant contribution to human well-being.

As noted by WBCSD, the challenge is to foster technological change that allows cultures to remain diverse and heterogeneous, while at the same time improving access to education and internet connectivity to ensure that people are 'more aware of the realities of their planet and everyone on it' (WBCSD 2010). The Vision 2050 project envisages changes in the concept of work, to include more part-time, flexi-time, tele-working, co-working and years off. Increased access to information should facilitate environmental monitoring and management. The impacts of other new technologies on biodiversity are less clear.

1.3.3 Linkages between trends and their relation to business and biodiversity

The projections summarized above outline some of the many factors that business must consider in order to prepare for and contribute to a more sustainable future. What are less obvious are the connections between these trends and biodiversity, and the implications of these linkages for business. This section considers a range of major trends that affect business today, assesses the linkages with biodiversity and ecosystem services, and identifies the business risks, implications and opportunities that may result.

We suggest that biodiversity loss and ecosystem decline are linked to various major trends affecting business, including social, economic and environmental changes (Figure 1.3). In most cases, the causality runs both ways: various factors influence the pace and scale of biodiversity loss; equally, the loss of biodiversity and ecosystems contributes to other major trends (MA 2005; UNEP 2007). In short, the business response to biodiversity loss cannot be defined in isolation from the business response to a range of major trends.

For example, the disturbance or conversion of coastal ecosystems – particularly mangrove forests and vegetated dunes – typically results in greenhouse gas emissions that contribute to climate change. Such removals may also exacerbate the severity of climate change impacts such as coastal flooding (Dahdouh-Guebas et al. 2005).

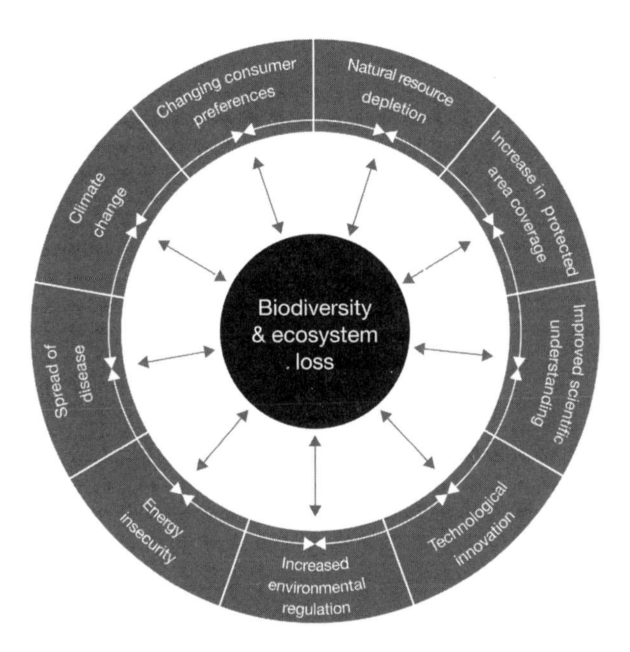

Figure 1.3 *Links between biodiversity and ecosystem loss and other major trends*

Source: PricewaterhouseCoopers for TEEB

Conversely, rising and increasingly stormy seas – which are among the expected impacts of climate change – can accelerate the loss of some coastal ecosystems (Sharpe 2000; Hulme and Jenkins 1998).

The linkages between biodiversity and other trends are explored further in Table 1.1, which examines a selection of major global trends in terms of how they affect potential biodiversity and ecosystem related risks and opportunities, and the implications for business. The table provides:

- a description of the trend and how it relates to biodiversity and ecosystem services;
- examples of the biodiversity-related risks this may present to business, along with potential business responses (see also Chapter 4); and
- examples of potential biodiversity-related business opportunities (see also Chapter 5).

The trends included in Table 1.1 are not exhaustive and the analysis is necessarily general, but this may provide a starting point for more detailed analysis along

Table 1.1 Major trends and their potential implications for biodiversity and business

Trend	Potential biodiversity-related business risks and implications	Potential biodiversity-related business opportunities
Natural resource depletion Diminishing supplies of raw materials and biological resources, such as fresh water, fertile soils, timber, fish, etc. This trend is exacerbated by pollution, climate change (see below), the spread of alien invasive species, and rising levels of consumption in many emerging economies. For example, in 2006 it was projected that the world's commercial fisheries will collapse in less than 50 years if current rates of fishing continue (Worm et al. 2006).	**Risk** • Increasing scarcity of natural resources implies reduced or more costly access. Secondary risks related to declining availability of natural resources (e.g. interstate conflict, resource nationalism, terrorism or mass migration) could further reduce business access. **Implication** • Business needs tools to monitor stocks of natural resources on which they rely, and to account for potential resource scarcity in long-term planning. • Business may need to find creative ways to secure access to their resource requirements, particularly fertile and well-watered agricultural land, that consider the needs of other stakeholders.	**Opportunity** • Resource efficiency will become more important to business competitiveness. Early adapters may gain competitive advantage. • Increasing scarcity of natural resources and resulting price increases should stimulate investment in resource-conserving and substitute technologies.

Table 1.1 Major trends and their potential implications for biodiversity and business *(Cont'd)*

Trend	Potential biodiversity-related business risks and implications	Potential biodiversity-related business opportunities
Increased protected area coverage Over the last three decades there has been a threefold increase in the total coverage of protected areas (UNEP 2007), and this expansion is expected to continue, particularly for under-represented marine and coastal areas.	**Risk** • Continued expansion of protected areas will restrict some business operations or increase operating costs for businesses that rely on access to, or conversion of, land/marine areas. This will especially influence sectors such as tourism, agriculture, forestry, fishing, shipping, mining and quarrying, oil and gas. **Implication** • Businesses will need to work at a regional or landscape level with peers, regulators and NGOs to secure their licence to operate. This may include direct business contributions to protected area objectives. • Some businesses may need to tighten their internal controls for environmental management in order to secure and maintain operating consent in and around protected areas.	**Opportunities** • Businesses able to generate the same output from a smaller land/sea 'footprint' will outperform their peers, where protected areas constrain access. • A track record of good environmental stewardship and support for protected areas may be viewed favourably by regulators when considering business requests for access to resources.
Improved scientific information A combination of research and improvements in information technology mean that ecological data is increasingly reliable, more easily accessible, and spatial data is of higher resolution. For example, 2007–2009 saw major quality and accuracy improvements in the ecological information on the World Database on Protected Areas, including the integration of marine protected areas, now available online (WDPA 2009).	**Risk** • Advances in the monitoring of natural resources will facilitate increased scrutiny by external stakeholders of business uses and impacts on biodiversity and ecosystems. **Implication** • Better evidence of how companies impact and depend upon biodiversity and ecosystem services will push environmental issues higher up the agenda of business priorities.	**Opportunity** • Companies that use Improved ecological information may gain advantage through earlier acquisitions of high value resources, ecosystem service agreements and/or operating licences.

Table 1.1 Major trends and their potential implications for biodiversity and business *(Cont'd)*

Trend	Potential biodiversity-related business risks and implications	Potential biodiversity-related business opportunities
Technological innovation Continued development of bio-mimicry engineering, bio-technology, etc. For example, in 2008, second generation bio-fuels (bio-chemical and thermo-chemical) reached the demonstration stage, while third generation algal bio-fuels promise further increases in productivity (IEA 2008).	**Risk** • Some new technologies and resource management systems may reduce genetic diversity (e.g. high-yield hybrid or monoclonal crop varieties) or impair ecosystems in other ways, generating both operational and reputational risks for business. **Implication** • More stringent safeguards to protect endangered species and ecosystems from the risks posed by new technology may be imposed, e.g. extended research and development procedures, increased quality control, bans on trials of new products close to sensitive habitats.	**Opportunity** • Business can use education and communication, in collaboration with peers and NGOs, to allay public concerns about new technology. • Potential commercial opportunities for businesses that invest in or develop new technologies and production practices that conserve biodiversity and ecosystem services.
Increasingly stringent environmental policy Public policies to protect nature and ensure that business pays for damage to biodiversity are increasingly stringent and enforced, e.g. EU Environmental Liability Directive, EU Habitats Directive, US Clean Water Act, Mexico's Sustainable Forestry Code (Ecosystem Marketplace 2010), Brazil's Compensação Ambiental Law.* Voluntary approaches or 'soft laws' are also increasingly demanding and often anticipate future regulatory reforms, e.g. biodiversity offsets, eco-certification and labelling.	**Risk** • Unforeseen regulatory change and increasing regulatory burden on business to reduce adverse impacts on biodiversity, with governments applying the 'polluter pays principle' more widely and stringently. • Compliance costs and 'green' taxes on carbon, water, land and other natural resources imply higher business costs. **Implication** • Businesses should monitor emerging environmental policy regimes and ensure they have procedures in place to identify, control and report their environmental performance. • More time and effort may be required for business expansion (e.g. permitting and planning consents, credit conditions) as potential impacts on biodiversity are subjected to greater scrutiny.	**Opportunity** • Some companies go 'beyond compliance' to prepare for impending regulatory change. Business may benefit by helping to shape future regulations and improve stakeholder relations. • Increased reliance of policy makers on market-based environmental policy, such as payments for ecosystem services, may offer new revenue opportunities for some businesses and/or make the mitigation of impacts more flexible and less costly.

Table 1.1 Major trends and their potential implications for biodiversity and business *(Cont'd)*

Trend	Potential biodiversity-related business risks and implications	Potential biodiversity-related business opportunities
Energy insecurity Diminishing and increasingly inaccessible fossil fuel reserves, combined with political risks to energy supply and concerns about climate change, are forcing countries and businesses to reassess and diversify their energy supplies.	**Risk** • Increasing reliance on land-intensive energy sources (bio-energy, concentrated solar thermal power, onshore wind, oil sands, etc.) further increases competition for and pressure on the land base. • Energy businesses increasingly look to technically challenging operating environments (e.g. deep water and arctic) to secure access to hydrocarbons. **Implication** • Businesses operating in areas where policy supports land-intensive energy generation will need to plan to ensure future access. For example, agribusiness may find it more difficult to secure fertile land, due to public policy support for the development of bio-fuels (e.g. in India, government policy requires that 20% of diesel fuel demand is met by bio-fuels by 2017, which may require up to 14 million hectares of land (NCAER 2009)).	**Opportunity** • Business may develop bio-crops or bio-fuel technologies that do not compete with food crops for land and water. • Opportunity to gain competitive advantage by planning ahead to secure future energy needs.
Spread of disease Changing patterns of disease and pandemics such as avian flu, swine flu and West Nile virus, exacerbated by poor water quality and other features of degraded ecosystems. Other trends such as climate change, urbanization and globalization may accelerate the spread of communicable diseases.	**Risk** • Society (hence business) relies on healthy ecosystems, including clean air and water to control the spread of diseases. Degraded ecosystems can jeopardize the health of consumers and employees, and affect business value chains. • Biodiversity loss could affect businesses that seek to exploit medicinal and other properties of wild plants and animals (e.g. in the health sector).	**Opportunity** • Spread of infectious diseases (exacerbated by poor water quality and degraded ecosystems) may lead to increased health spending and could provide increasing opportunities for the healthcare sector. • Business can develop tailored treatments or technologies to mitigate and/or adapt to the health consequences of ecosystem decline.

Table 1.1 Major trends and their potential implications for biodiversity and business *(Cont'd)*

Trend	Potential biodiversity-related business risks and implications	Potential biodiversity-related business opportunities
	Implication • Businesses can assess how their operations would be affected by disease epidemics, and take actions to reduce the spread of disease among employees. • Businesses that rely on wild genetic resources (e.g. biotechnology, pharmaceuticals) should plan for a world of declining biodiversity and more stringent benefit-sharing requirements.	
Climate change Complex phenomena attributed to greenhouse gas emissions are changing the functioning of ecosystems at regional and global levels. Under various scenarios, the IPCC suggests a global temperature increase of between 1 and 6 °C by the end of this century (IEA 2009). Even with a global average temperature increase of just 1.5–2.5 °C, it is expected that 20–30% of all species are likely to be at increased risk of extinction (IPCC 2007).	**Risk** • Changing temperatures, increased extreme weather events, sea-level rise, increased water stress and drought will dramatically alter the availability of ecosystem services upon which all businesses rely. For example, loss of natural tourism assets such as coral reefs due to changes in sea temperature and acidity, or reduction of agricultural yields due to increased water scarcity. **Implication** • Businesses can integrate climate change impacts into long-term planning and assess where this trend may jeopardize access to ecosystem services.	**Opportunity** • Development of business services and tools to evaluate risks associated with climate change (e.g. climate risk mapping) or provide climate adaptation services (e.g. drought-resistant crops). • Potential advantage for businesses that can anticipate climate change impacts and 'climate proof' their business models. • Participation in emerging markets for bio-carbon offsets (including REDD+).
Rise of responsible finance Responsible investing now accounts for more than 7% of global assets under management and this is projected to grow to 15–20% by 2015 (Robecco and Booz & Company 2008). Investment criteria increasingly include a range of environmental indicators. Corporate lenders are also tightening requirements for the provision of finance and increasingly considering the biodiversity impacts of prospective clients or projects.	**Risk** • Restricted access or increased costs of finance for companies which have adverse impacts on biodiversity and ecosystems, or cannot show that they are taking appropriate actions to avoid, mitigate or compensate for such impacts.	**Opportunity** • Cheaper and easier access to finance for companies with demonstrable positive biodiversity impacts may be a source of competitive advantage. • Opportunity for data providers to service the growing need for robust spatial biodiversity and protected area datasets.

Table 1.1 Major trends and their potential implications for biodiversity and business *(Cont'd)*

Trend	Potential biodiversity-related business risks and implications	Potential biodiversity-related business opportunities
Recent revisions to the International Finance Corporation's Performance Standard 6 (covering the biodiversity impacts of project finance) will reinforce this trend and may further restrict access to finance for projects in sensitive areas.	**Implication** • Businesses operating in sectors or areas with the potential for significant adverse impacts on biodiversity and ecosystems may need to prove to prospective financiers that their management practices are sufficient to avoid or offset negative impacts on ecosystems or that their areas of operation do not affect protected or ecologically sensitive sites. • Businesses and their financiers will need improved information on the spatial distribution of protected areas and other ecologically sensitive sites.	
Changing consumer preferences		
Section 1.3.4 provides a more detailed exploration of the impacts and implications for business of changes in consumer preferences with respect to biodiversity and ecosystems. This trend is singled out for more detailed treatment because it may have far-reaching impacts on a range of business value chains, and also as an illustration of the potential for more detailed analysis of other trends.		

Note: * The 'Compensação Ambiental' is described within Article 36 of Brazilian law (Law nr. 9985) and is designed to offset the negative impacts on the natural environment from project development, requiring developers to pay a licensing fee.

company or industry lines. The list of trends has not been prioritized, and in fact the relevance of each trend will vary depending on a company's geographic exposure and business activities. Further detail is given in the following section on trends in consumer preference, pointing the way to deeper analysis of the links between major global trends, biodiversity and business.

Some of the trends outlined in Table 1.1 lie outside the traditional sphere and mandate of business sustainability, environmental or biodiversity management systems. However, as biodiversity and ecosystems are linked to many other trends, they should not be considered in isolation. Business risk management systems can help make the connections and provide a structure to analyse and track such trends, allocate resources and determine responses. We suggest that if biodiversity and ecosystem risks are identified, assessed and managed early, they can provide a basis for business competitive advantage.

1.3.4 Changing consumer preferences: Implications for business and biodiversity

Business, biodiversity and the linkages between them are heavily influenced by consumer preferences, which are constantly evolving. A recent survey of over 13,000 people suggests that consumers are more concerned about the environment today than they were just a few years ago: 82 per cent of Latin American consumers were more concerned; 56 per cent in Asia; 49 per cent in the USA; and 48 per cent in Europe (Taylor Nelson Sofres 2008). Some examples of changing consumer preferences include reduced demand for traditional Chinese medicine, due to perceived impacts on endangered species (e.g. tigers, bears, sea horses), or the changing acceptance of fur clothing in Europe and North America, with knock-on effects on both wild hunting and farming of animals for their fur.

Public awareness of biodiversity is also growing: a 2010 survey by IPSOS for the Union for Ethical BioTrade (UEBT), published as the 'Biodiversity Barometer', revealed that 60 per cent of consumers in Europe and the United States, and 94 per cent in Brazil, had heard of biodiversity, representing a significant increase on the previous year (UEBT 2010). Increasing awareness is likely to influence purchasing behaviour, with 81 per cent of consumers interviewed in the same survey declaring that they would cease buying products from companies that disregard ethical sourcing practices. In another survey of UK consumers, conducted in May 2010, around half of all respondents indicated that they would be willing to pay between 10 and 25 per cent more for purchases up to GB£100, in order to account for their impacts on biodiversity and ecosystems.[2]

The proliferation of ecologically certified products is another indication of changing consumer preferences: the UEBT Biodiversity Barometer referred to above also revealed that 82 per cent of consumers would have more faith in companies that subject themselves to independent verification of their sourcing practices (UEBT 2010). Many labelling schemes arose in response to NGO campaigns, public concerns and changing preferences related to biodiversity loss, including the Forest Stewardship Council (FSC), Marine Stewardship Council (MSC) and Rainforest Alliance certified coffee, cocoa and tea, while membership of the International Social and Environmental Accreditation and Labelling Alliance has more than doubled over the last two years (ISEAL 2009).

In addition to the increased number of eco-labelling schemes, total sales and the market share of certified products are also growing, albeit from a small base. Between 2005 and 2007, for instance, sales of FSC-labelled goods quadrupled (FSC 2008), while spending on ethical food and drink in general has increased more than threefold over the last decade, growing from GB£1.9 billion in 1999 to over GB£6 billion in 2008 (Co-operative Bank 2008). In another example, between April 2008 and March 2009 the global market for MSC-labelled seafood products grew by over 50 per cent to reach a retail value of US$1.5 billion (MSC 2009).

The behaviour of some FMCG (fast-moving consumer goods) brand owners suggests that eco-labelling is moving from niche markets into the mainstream. In recent years, several brand owners and retailers have added ecologically friendly product attributes to their major brands, often through certification. Examples include Domtar (FSC certified paper), Mars (Rainforest Alliance cocoa), Cadbury (Fairtrade cocoa), Kraft (Rainforest Alliance Kenco coffee) and Unilever (Rainforest Alliance PG Tips). Importantly, all these brands offer biodiversity attributes through certification schemes but do not ask consumers to pay a premium or to compromise on quality, taste or availability. Retailers are also taking action on biodiversity and communicating that action to consumers. In the UK, for example, the Waitrose supermarket chain links its palm oil policy to customer labelling: 'Waitrose already has in place a technical policy to name oils, rather than use the term "blended vegetable oils". As a result we can confirm that palm oil is used as an ingredient in only a small number of our own branded products, which are identifiable to our customers' (Waitrose 2009).

Business action is not only responding to changing consumer preferences but is itself a key driver in influencing and educating consumers. Governments can also influence consumer choice and producer behaviour by market regulation and incentives (e.g. taxes and subsidies), but also through their own purchasing strategies. For example, 16 EU Member States have adopted Green Public Procurement (GPP) National Action Plans, which include environmental criteria for the purchase of products and services (EC 2009).

As consumer demand for biodiversity-friendly products and services increases, businesses should attempt to ensure that the implications for their operations are identified, assessed and managed. Increasingly, the supply chains of consumer goods companies are under scrutiny in terms of their biodiversity impacts and management. For example, Walmart has begun scoring its suppliers based on their sustainability performance, with biodiversity and use of natural resources featuring heavily in the process. Walmart aims to establish eco-labels for all its products within five years. Pressure on suppliers to embed biodiversity considerations within their internal management processes is thus likely to increase.

For consumer-facing businesses, in particular, companies may wish to ensure that biodiversity is fully embedded in risk management systems. This may include:

- Ensuring that biodiversity-related consumer concerns are included on the corporate risk register: how well does the business know its customers' attitudes?
- Assessing significance of the risk: how will it impact brand value? Are biodiversity-related shifts in customer preferences likely to influence demand for key products? How does this relate to other risks such as climate change, water scarcity, business cycles?
- Designing appropriate responses: modify internal procurement or production procedures and influence key players in the value chain to ensure that biodiversity impacts are minimized, establish collaborative processes to combine expertise across an industry on biodiversity-related consumer issues, devise specific biodiversity policy and communications strategy to address customer concerns and educate consumers.

In order to capitalize on growing markets for environmentally responsible goods and services, businesses may wish to consider whether they have relevant processes and expertise. For example, adherence to eco-certification schemes requires a thorough understanding of the biodiversity impacts of business products and processes, as well as the development of capacity to support monitoring, system controls, evaluation and reporting systems.

1.4 From major trends to business values

The preceding section reviewed some major trends that affect business and showed how these trends may influence business responses to biodiversity loss and ecosystem decline.

While many of these trends lie outside the traditional sphere of biodiversity and ecosystem management, we argue here that they cannot be considered in isolation.

In order to develop effective responses to these trends and their interactions, business needs reliable information to assess its impacts and dependence on biodiversity and ecosystems. Risk management frameworks can provide a structure and process for analysing and tracking such trends, allocating resources and determining adequate responses. If significant risks are identified, assessed and managed early, they can be transformed into competitive advantage. The next chapter of this book provides an overview of the status and trends in biodiversity and ecosystems, the drivers of biodiversity loss, and describes the economic values at stake. The impact and dependencies of a range of sectors on BES are also explored, together with an overview of how these create both risks and opportunities for business.

Acknowledgements

Annika Andersson (Vattenfall), Celine Tilly (Eiffage), Christoph Schröter-Schlaack (UFZ), Daniel Skambracks (KfW Bankengruppe), Deric Quaile (Shell), Dorothea Seebode (Philips), Elaine Dorward-King (Rio Tinto), Gemma Holmes (PricewaterhouseCoopers), Gérard Bos (Holcim), Jennifer McLin (IUCN), Juan Gonzalez-Valero (Syngenta), Juan Marco Alvarez (IUCN), Jun Hangai (Nippon Keidanren), Kerstin Sobania (TUI), Kii Hiyashi (Nagoya University), Lloyd Timberlake (WBCSD), Margaret Adey (Cambridge U.), Mônica Barcellos (UNEP–WCMC), Naoki Adachi (Response Ability), Nina Springer (Exxon/IPIECA), Oliver Schelske (SwissRe), Olivier Vilaca (WBCSD), Paul Hohnen, Per Sandberg (WBCSD), Polly Courtice (Cambridge U.), Ravi Sharma (CBD Sec.), Roberto Bossi (ENI), Ruth Romer (IPIECA), Ryo Kohsaka (Nagoya City U.), Sachin Kapila (Shell), Sagarika Chatterjee (F&C Investments), Simon Anthony, Toby Croucher (Repsol/IPIECA), Valerie David (Eiffage), Virpi Stucki (IUCN).

Notes

1 More information about this book and other TEEB outputs can be found at www.teebweb.org.

2 Survey carried out on behalf of PricewaterhouseCoopers by Opinium, in May 2010, with over 2,000 respondents across the UK. Respondents were asked a mix of single and multiple-choice questions.

References

Athanas, A., Bishop, J., Cassara, A., Donaubauer, P., Perceval, C., Rafiq, M., et al. (2006) *Ecosystem Challenges and Business Implications*. Business and Ecosystems Issue Brief, Earthwatch Institute, World Resources Institute, WBCSD and IUCN (November).

Bishop, J., Kapila, S., Hicks, F., Mitchell, P. and Vorhies, F. (2008) *Building Biodiversity Business*. Shell International Ltd and IUCN: London, UK, and Gland, Switzerland. 164 pp. (March).

Co-operative Bank (2008) *Ten Years of Ethical Consumerism, 1999–2008*. URL: www.ethicalconsumer.org/Portals/0/Downloads/ETHICAL%20CONSUMER%20REPORT.pdf (last accessed 9 October 2009).

Dahdouh-Guebas, F., Jayatissa, L. P., D. Nitto, D., Bosire, J.O., Lo Seen, D. and Koedam, N. (2005) 'How effective were mangroves as a defence against the recent tsunami?' *Current Biology* vol. 15, no. 12. URL: www.vub.ac.be/APNA/staff/FDG/pub/Dahdouh-Guebasetal_2005b_CurrBiol.pdf (last accessed 17 June 2010).

EC (2009) *National GPP Policies and Guidelines*. URL: http://ec.europa.eu/environment/gpp/national_gpp_strategies_en.htm (last accessed 9 October 2009).

Ecosystem Marketplace (2010) *State of Biodiversity Markets: Offset and Compensation Programs Worldwide*. Forest Trends, Washington, DC.

FSC (2008) 'Facts and figures on FSC growth and markets'. URL: www.fsc.org/fileadmin/web-data/public/document_center/powerpoints_graphs/facts_figures/2008-01-01_FSC_market_info_pack_-_FINAL.pdf (last accessed 9 January 2009).

Hulme, M. and Jenkins, G.J. (1998) *Climate Change Scenarios for the UK; Scientific Report*. UKCIP Technical Report No. 1, Climatic Research Unit, Norwich, UK.

IEA (2009) *World Energy Outlook*. OECD, Paris.

IEA (2008) *From 1st–2nd Biofuel Generation Technologies*. URL: www.iea.org/papers/2008/2nd_Biofuel_Gen_Exec_Sum.pdf (last accessed 9 January 2010).

IPCC (2007) *Fourth Assessment Report Climate Change, Synthesis Report*. URL: www.ipcc.ch/ (last accessed 9 January 2010).

ISEAL (2009) pers. comm., September.

MA (2005) *Ecosystems and Human Well-being: Opportunities and Challenges for Business and Industry*. Island Press, Washington, DC.

McKinsey & Company (2007) *How Companies Think about Climate Change: A McKinsey Global Survey*. McKinsey & Company, www.mckinseyquarterly.com.

McKinsey & Company (2010) *The Next Environmental Issue for Business: McKinsey Global Survey results*. McKinsey & Company, www.mckinseyquarterly.com.

MSC (2009) Annual Report 2008/2009. URL: www.msc.org/ (last accessed 9 October 2009).

NCAER (2009) *Biodiesel from Jatropha: Can India Meet the 20% Blending Target?* Elsevier. URL: http://linkinghub.elsevier.com/retrieve/pii/S0301421509008593 (last accessed 9 January 2010).

Nippon Keidanren (2010) *Declaration of Biodiversity*, URL: www.keidanren.or.jp/english/policy/2009/026.html (last accessed 15 June 2010).

PricewaterhouseCoopers (2010) *13th Annual Global CEO Survey 2010*. Available at: www.pwc.com/gx/en/ceo-survey/index.jhtml (last accessed 15 June 2010).

Robecco and Booz & Company (2008) *Responsible Investing: A Paradigm Shift from Niche to Mainstream*.

Sharpe, J. (2000) *Coast in Crisis, Protecting Wildlife from Sea Level Rise and Climate Change, Royal Society* for the Protection of Birds, UK. URL: www.rspb.org.uk/Images/CRISIS72_tcm9-133013.pdf (last accessed 17 June 2010).

Taylor Nelson Sofres (2008) Global Shades of Green – TNS Green Life Study, presented at TNS Green Life Conference in New York City, October.

TEEB Foundations (2010) *The Economics of Ecosystems and Biodiversity: Ecological and Economic Foundations* (ed. P. Kumar), Earthscan, London.

UEBT (2010) Biodiversity Barometer 2010 URL: www.countdown2010.net/2010/wpcontent/uploads/UEBT_BIODIVERSITY_BAROMETER_web-1.pdf (last accessed 25 May 2010).

United Nations (1993) Convention on Biological Diversity (with annexes). Concluded at Rio de Janeiro on 5 June 1992. Treaty series No. 30619. URL: www.cbd.int/convention/convention.shtml

UNEP (2007) Global Environment Outlook: Environment for Development. GEO4. UNEP/Earthprint. URL: www.unep.org/geo/geo4/media/ (last accessed 19 May 2010).

Waitrose (2009) 'Palm oil policy'. URL: www.waitrose.com/food/foodissuesandpolicies/palmoil.aspx (last accessed 9 October 2009).

WBCSD (2010) Vision 2050: The New Agenda for Business. World Business Council for Sustainable Development, Geneva (February). Available at: www.wbcsd.org/web/vision 2050.htm

WDPA (2009) World Database on Protected Areas Annual Release. URL: www.wdpa.org/AnnualRelease.aspx (last accessed 9 January 2010).

Worm, B., Barbier, E., Beaumont, N., Duffy, E., Folke, C., Halpern, B., et al. (2006) Impacts of Biodiversity Loss on Ocean Ecosystem Services, *Science* vol. 314. no. 5800, pp. 787–790. URL: www.sciencemag.org/cgi/content/abstract/314/5800/787 (last accessed 9 November 2010).

Chapter 2

Business Impacts and Dependence on Biodiversity and Ecosystem Services

Editors
Mikkel Kallesoe (WBCSD), Nicolas Bertrand (UNEP)

Contributing authors
Scott Harrison (BC Hydro), Kathleen Gardiner (Suncor Energy Inc.), Peter Sutherland (GHD), Bambi Semroc (CI), Julie Gorte (Pax World), Eduardo Escobedo (UNCTAD), Mark Trevitt (Trucost plc), Nathalie Olsen (IUCN), James Spurgeon (ERM), John Finisdore (WRI), Jeff Peters (Syngenta AG), Ivo Mulder (UNEP FI), Christoph Schröter-Schlaack (UFZ), Emma Duncan, Cornelia Iliescu (UNEP), Annelisa Grigg (Global Balance)

Contents

Key messages

All companies, regardless of sector, both impact on biodiversity and ecosystems and depend on ecosystem services. It is hard to think of any economic activity that does not benefit from biodiversity and ecosystem services (BES) or in some way modify the ecosystems around it. For example, the bio-tech industry benefits from access to wild genetic resources, but may also create risks through the introduction of genetically modified organisms; agribusiness and the food sector depend on ecosystem services like pollination, but also, through impacts on land and water resources, reduce other ecosystem services; forest industries, construction and publishing rely on sustained supplies of timber and wood fibre, but can alter forest structure at the expense of wildlife and recreational values; and tourism derives profit from the cultural services and aesthetic values of natural landscapes but may bring so many tourists into an area that nature values are diminished.

BES decline continues at unprecedented rates. Most indicators of the state of BES show declines, while indicators of pressures on biodiversity show increases, and, despite some local successes and responses, the rate of biodiversity and ecosystem loss does not appear to be slowing. This poses real and tangible risks for business and for society in general, as BES generate value for businesses and the wider economy, and the loss of BES imposes both private and public costs.

Principal pressures directly driving the decline in BES are habitat change, over-exploitation, pollution, invasive alien species and climate change. Businesses can help reduce these pressures by managing and mitigating their impacts on biodiversity and ecosystem services. They should systematically review their operations in relation to BES and assess how direct and indirect drivers of change in ecosystem services may affect their business.

Operational, regulatory, reputational, market, product and financial risks associated with BES decline are often overlooked and underestimated by business, particularly when they are indirect. Companies need to examine their entire value chain in order to determine how and where BES impacts and dependence may affect their business. Although historically BES has been given little attention in financial analyses of company performance, this is changing, partly as a consequence of increased attention to climate change risks and opportunities in business.

There are untapped business opportunities to address BES decline while contributing to other societal goals. Far-sighted businesses can create opportunities from the greening of investor, client and consumer preferences. However, businesses that fail to assess their impacts and dependence on BES may neglect some of these profitable opportunities.

The values of BES are often external to business decision making. While acknowledging the importance of BES decline and measuring ecosystem impacts and dependencies, many companies still find it difficult to integrate such information into their core operational and corporate decisions.

2.1 Introduction

Most companies have a two-way relationship with nature. On the one hand, they may have direct impacts on biodiversity and ecosystems through their core operations, indirectly through their supply chain, or through their lending and investment choices. On the other hand, many companies depend on biodiversity and the services provided by ecosystems as key inputs to products and production processes.

It is hard to think of any economic activity that does not benefit from ecosystem services in some way (Hanson et al. 2008). Fresh water, for instance, is a critical input to almost every industrial process – from fresh-cut lettuce processing to large-scale mining. Pharmaceutical companies rely on wild genetic resources to identify new active compounds. Agribusiness depends on natural pollination, pest control and soil biological processes. Many tourism destinations owe their attractiveness to surrounding natural environments. The protection offered by ecosystems such as wetlands and mangroves can reduce damage from storms and flooding, and is monitored by insurance and reinsurance companies.

How a business conducts its operations can affect the value of biodiversity generally or a specific ecosystem service – for the company itself, as well as for other sectors and society as a whole. Today, however, most company managers pay insufficient attention to the links between biodiversity, ecosystems and their business. While some companies recognize the importance of their impacts and dependence on biodiversity and ecosystem services, many others struggle to understand how to integrate such information into day-to-day business.

The Millennium Ecosystem Assessment (MA) offers a sobering account of biodiversity loss and ecosystem degradation (MA 2005a). Although this loss of natural capital is still not adequately reflected in national economic statistics or company accounts, the impacts are real and increasingly seen as material by businesses.

New public policies and regulations are being developed in response to biodiversity loss and ecosystem degradation – at global, regional and local levels. Business impacts on BES are under increasing scrutiny from customers, investors, employees and regulators. More and more business managers are taking action to improve their understanding and manage their biodiversity and ecosystem impacts and dependence, while also developing new business solutions to meet these challenges.

This chapter summarizes how the status, trends and forecasts for biodiversity, ecosystems and ecosystem services are affecting business. We present the main drivers of biodiversity loss and ecosystem degradation, introduce the concept of externalities and describe the economic values at stake. We highlight typical impacts and dependencies on biodiversity and ecosystems for a range of business sectors, and outline how these create both risks and opportunities.

2.2 Biodiversity, ecosystems and ecosystem services

Biodiversity is defined by the Convention on Biological Diversity (CBD) as the variability among living organisms within species (genetic variation), between species and between ecosystems. Species richness is probably the best documented of these three components. Today, approximately 1.75 million species are known to science, although plausible estimates suggest that the total number of species on Earth ranges from 5 million to 30 million.

Ecosystems are a major component of biodiversity and are defined by the MA as a dynamic complex of plant, animal and micro-organism communities and the non-living environment interacting as a functional unit (MA 2005a). Examples include deserts, coral reefs, wetlands, rainforests, boreal forests, grasslands, urban parks and cultivated farmland.

'Ecosystem services' are the benefits that people obtain from ecosystems. These are sometimes grouped into four categories (Table 2.1).

The value of ecosystem services is intimately linked to biodiversity. Table 2.2 illustrates how the ecosystem services enjoyed by people depend on both the diversity (quality) as well as the sheer amount (quantity) of genes, species and ecosystems found in nature.

2.2.1 Biodiversity, ecosystems and ecosystem status and trends

Over the past 50 years, humans have altered ecosystems more than in any comparable period in our history, largely in order to meet growing demand for food, fresh water, timber, fibre, energy and other materials.

While the use of natural resources has helped to satisfy human needs, an unintended consequence has been widespread fragmentation, degradation or outright conversion

Table 2.1 Four categories of ecosystem services

Provisioning	Goods or products, such as food, freshwater, timber and fibre
Regulating	Benefits from natural processes, such as climate regulation, disease and pest control, soil formation and stabilization, water filtration, pollination, as well as protection from floods, storm surges and other natural hazards. Note that 'regulating' in this context is a natural phenomenon and should not be confused with government policies or regulations.
Cultural	Non-material benefits, such as recreation, spiritual inspiration and aesthetic enjoyment of nature.
Supporting	Fundamental natural processes, such as nutrient cycling and primary production, which underpin all other ecosystem services.

Source: Adapted from MA (2005c)

Table 2.2 Relationship between biodiversity, ecosystems and ecosystem services

Biodiversity	'Quality'	'Quantity'	Services (examples)
Ecosystems	Variety	Extent	• Recreation • Water regulation • Carbon storage
Species	Diversity	Population	• Food, fibre, fuel • Design inspiration • Pollination
Genes	Variability	Number	• Medicinal discovery • Disease resistance • Adaptive capacity

of ecosystems, leading to the loss of biodiversity and reductions in the quality and quantity of important ecosystem services. A key indicator of ecological deterioration is the risk that an increasing number of species may become extinct, as well as the genetic impoverishment of remaining populations.

Species extinction is a natural part of the evolutionary process. However, the rate of species loss in recent decades is estimated to be 100–1,000 times faster than the 'natural' rate. The largest declines are thought to have occurred in temperate and tropical grasslands and forests, areas where human civilizations first developed and disturbance seems most pronounced.

Amphibians face the greatest risk of extinction and coral species are deteriorating most rapidly (CBD 2010). Furthermore, the abundance of vertebrate species (based on assessed populations) fell by nearly one-third on average between 1970 and 2006, and continues to fall globally, with especially severe declines in the tropics and among freshwater species (CBD 2010). Other assessments show similar deterioration across a range of indicators (Butchart et al. 2010).

The loss of a single species can have wider effects on other species and entire ecosystems. In general, species extinction appears to reduce the resilience of ecosystems, leaving them at greater risk of further deterioration. Recent projections of the impacts of climate change indicate continuing and accelerating rates of species extinction, continued loss of natural habitats, and changes in the distribution and abundance of species, species groups and biomes (CBD 2010).

At the level of ecosystems, primary forests have completely disappeared in a number of countries and, every year, millions of hectares are lost to deforestation, mainly in Latin America, Southeast Asia and Africa (FAO 2001). The world has lost roughly half its wetlands since 1900 (UNWWAP 2003), and about 20 per cent of its mangrove forests between 1980 and 2005 (FAO 2007). In addition, 20 per cent of the world's coral reefs have been destroyed (MA 2005a) and a further 30 per cent have been seriously damaged by destructive fishing practices, pollution, disease, coral bleaching, invasive alien species and unsustainable tourism (Wilkinson 2008).

Such rapid alterations are compromising the ability of ecosystems to recover from extreme events and external shocks. Evidence is growing that many species and ecosystems are nearing their 'tipping points', where further disturbance may result in abrupt and possibly irreversible decline in the benefits they provide. GBO-3 identified the following tipping points:

- the dieback of large areas of the Amazon forest, due to interactions between climate change, deforestation and fires, with adverse impacts on the global climate, regional rainfall patterns and species survival;
- changes in the chemistry of many freshwater lakes and other inland water bodies due to runoff from agriculture, industry and urban areas, leading to the build-up of nutrients, algae blooms, fish die-off and reduced recreational value; and
- collapsing coral reef ecosystems due to a combination of climate change (which results in ocean acidification, warmer water and coral bleaching), as well as overfishing, destructive fishing practices and nutrient pollution.

The impacts of changes in ecosystem extent and quality, species diversity and abundance, and genetic variety can be expressed in terms of the loss or decline of ecosystem services. According to the MA, almost two-thirds of the 24 ecosystem services they examined have been significantly reduced over the past 50 years, including almost all non-commodity benefits (Figure 2.1).

2.2.2 Modelling the future of biodiversity and ecosystem services

If we continue on our current development path and maintain today's patterns of resource exploitation, the world is likely to continue to lose biodiversity and many ecosystem services will be further reduced. However, it is difficult to predict the exact rates at which biodiversity and ecosystem change will occur.

The MA developed four scenarios for the period 2000–2050. All four scenarios indicate a general increase in provisioning services, achieved primarily through land-use change, but at the cost of further degradation in supporting, regulating and cultural services. Additional common results across the four scenarios include:

- increased demand for provisioning services;
- low levels of food security and child nutrition;
- fundamental modifications of freshwater resources;
- growing demand for fish and fish products leading to increased risk of decline in regional marine fisheries, with aquaculture unable to alleviate pressures due to its dependence on marine fish as feed;
- provision of ecosystem services is driven to a large extent by land-use change;
- provision of clean water is further impaired by wetland drainage and conversion;
- uncertainty over the role of terrestrial ecosystems as a net CO_2 sink; and
- difficult trade-offs between food and water supply.

Balance sheet: Ecosystem services

Provisioning services		
Food	crops	↑
	livestock	↑
	capture fisheries	↓
	aquaculture	↑
	wild foods	↓
Fibre	timber	+/–
	cotton, silk	+/–
	wood fuel	↓
Genetic resources		↓
Biochemicals, medicines		↓
Water	freshwater	↓

Regulating services	
Air quality regulation	↓
Climate regulation – global	↑
Climate regulation – regional and local	↓
Water regulation	+/–
Erosion regulation	↓
Water purification and waste treatment	↓
Disease regulation	+/–
Pest regulation	↓
Pollination	↓
Natural hazard regulation	↓
Cultural services	
Spiritual and religious values	↓
Aesthetic values	↓
Recreation and ecotourism	+/–

↑ globally enhanced
↓ globally degraded

The MA evaluated the global status of provisioning, regulating and cultural services. An upwards arrow indicates that the condition of the service globally has been enhanced and a downwards arrow that it has been degraded in the recent past.

Figure 2.1 *The ecosystem services balance sheet*

Source: Adapted from Millennium Ecosystem Assessment (2005a)

Predictions by the OECD suggest that agriculture will continue to be a major source of pressure on biodiversity; they highlight the risk of a business-as-usual scenario that would result in the loss of additional mature forests around the world by 2030: 68 per cent in south Asia, 26 per cent in China, 24 per cent in Africa and about 20 per cent in eastern Europe, Australia and New Zealand (OECD 2008).

The MA scenarios imply both risks and opportunities for business (MA 2005b). New markets may emerge for companies that are able to meet increased future demand for food, fibre and fresh water in a sustainable way. On the other hand, businesses involved in fish products and related activities will face increasing challenges, unless new technologies or fisheries management practices are adopted that can sustain productivity. Wetland conservation is expected to become a major public concern that businesses will need to factor into planning and decision making. Finally, carbon capture and storage technologies, as well as ecosystem conservation and restoration, may become significant business opportunities as part of climate change mitigation and adaptation strategies.

2.2.3 Drivers of biodiversity loss and ecosystem degradation

Effective responses to biodiversity loss and ecosystem decline begin with understanding the causes of environmental change. The various factors that directly or indirectly result in the loss of biodiversity are known as drivers or pressures. So-called direct drivers unequivocally influence biodiversity and ecosystems, but often vary in their importance depending on the context. Direct drivers (or pressures) include land conversion, over-exploitation, nutrient deposition, diseases and invasive alien species, and climate change. These affect different ecosystems in different ways (Figure 2.2).

Indirect drivers operate more diffusely, by accelerating one or more direct drivers. The MA identifies several indirect drivers, namely demographic, economic, socio-political, scientific and technological change, as well as cultural and religious trends (MA 2005c). Changes in biodiversity and ecosystems are almost always the result of the interaction of multiple drivers acting across different spatial, temporal and organizational scales (Figure 2.3).

Over the past 50 years, the most significant direct driver for terrestrial ecosystems has been land-use change, mostly due to agricultural expansion and urbanization. The latter includes both direct expansion of urban areas and indirect impacts due to the development of transport and infrastructure networks.

In marine ecosystems, the most important direct driver of change over the same period has been industrial fishing. Technical advances in the fishing industry, including more and larger boats, and more efficient fishing equipment, have depleted fish stocks to such a degree that it has affected the overall biodiversity, structure and functioning of the oceans. Pressure from fishing on some marine ecosystems is now so intense that commercial fish stocks have been reduced by up to 90 per cent compared with levels that existed prior to the onset of industrial fishing. A classic example is the collapse of the Atlantic cod stocks off the east coast of Newfoundland in 1992 (MA 2005a).

For freshwater ecosystems, the most important direct drivers of change over the past 50 years vary across regions, but include the modification of hydrological regimes, invasive alien species and pollution. As mentioned above, about 50 per cent of the world's wetlands have been lost since 1900 and invasive species are now a major cause of species extinction in freshwater systems. Discharge of nutrients from

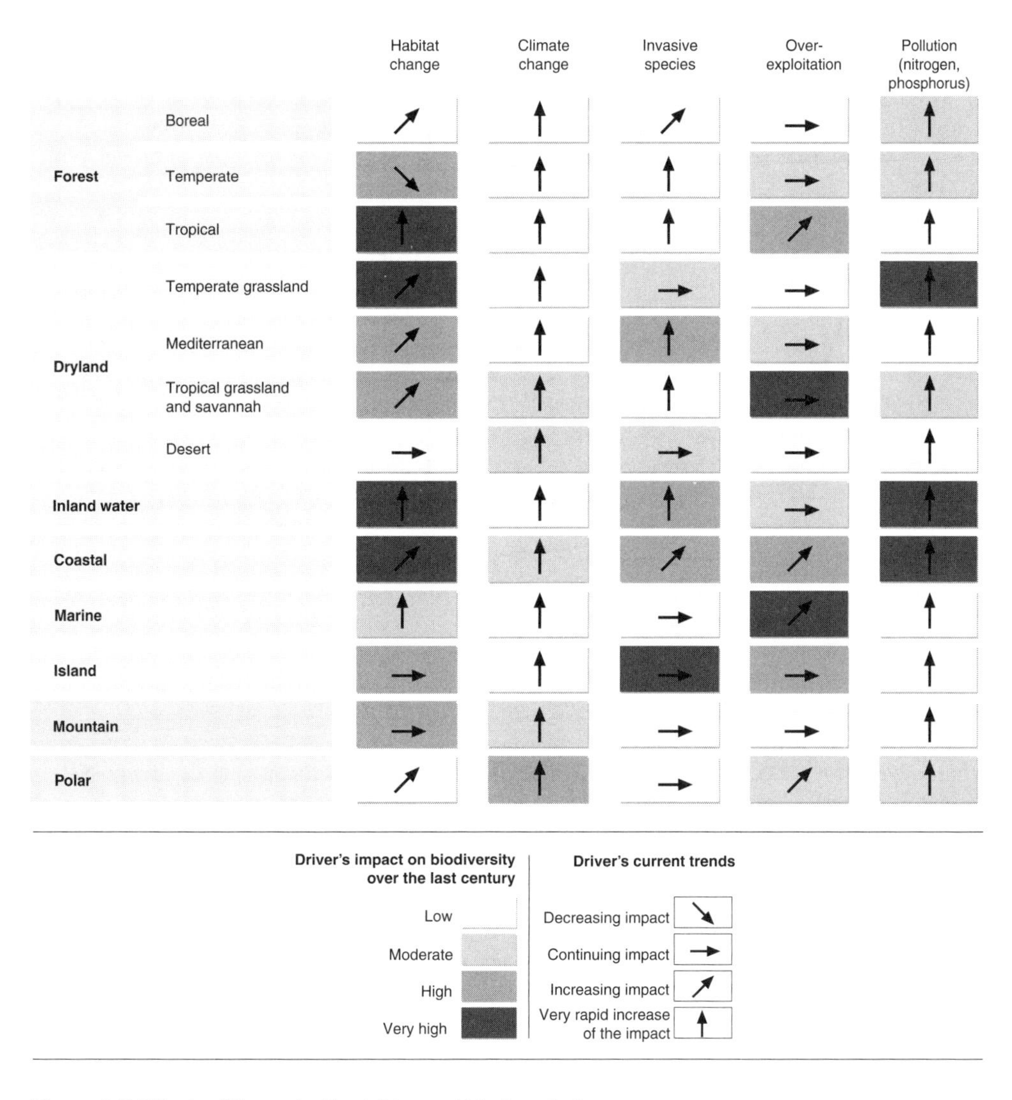

Figure 2.2 *Effects of the main direct drivers of biodiversity loss*

Source: Millennium Ecosystem Assessment (2005a)

agriculture, industry and built areas has caused widespread eutrophication and increased levels of nitrates in drinking water, while pollution from point sources (such as mining) has had significant adverse impacts on the biodiversity of some inland waters (MA 2005c).

In many cases, actions taken to enhance one ecosystem service will exacerbate pressures on other services. For example, increased food production has typically resulted in the loss of other ecosystem services, through impacts on land cover, increased fresh water withdrawals for irrigation and release of agro-chemical nutrients into surface water.

A more recent assessment indicates that the main direct drivers of biodiversity loss and ecosystem degradation are either constant or increasing in intensity (CBD 2010). Moreover, today many drivers converge in the same place and at a greater intensity

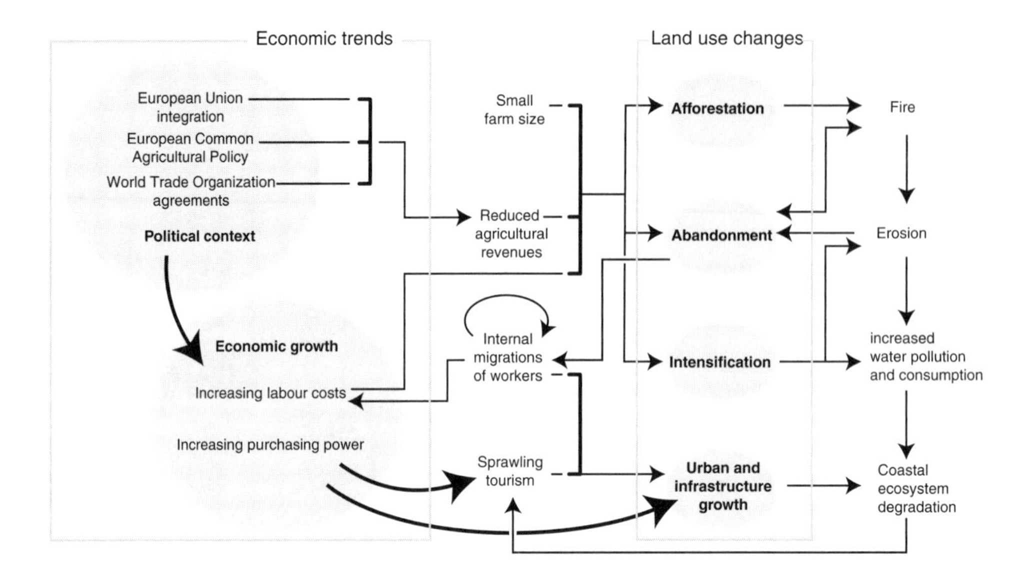

Figure 2.3 *Feedbacks and interaction between drivers*

Source: Millennium Ecosystem Assessment (2005a)

than ever before. Because exposure to one pressure can make a species or an ecosystem more susceptible to a second, and so on, multiple pressures can accumulate with dramatic impacts on biodiversity and ecosystems.

2.2.4 Implications for business

The MA identified six ecosystem challenges of particular concern for business, which are each examined below.

Fresh water scarcity

Forest and mountain ecosystems are the source of fresh water for two-thirds of the world's population. The quantity of available fresh water per person varies globally, but only about 15 per cent of the world's population enjoys abundant supplies of fresh water. Currently, between 1 and 2 billion people lack access to sufficient clean water to meet their needs, affecting food production, human health and economic development.

Most businesses depend on reliable sources of water for their operations. Many businesses also influence water quality through their wastewater discharge. Overuse of fertilizer, poor sanitation facilities and storm water runoff are other causes of declining fresh water quality.

Fresh water scarcity creates both business risks and opportunities. Risks include increased water costs, unpredictable water supply, government imposed water restrictions or rationing and reputational damage as a result of poor water treatment or inefficient use. Opportunities for business include improving the efficiency of water use through market mechanisms (e.g. water trading) or new technologies (e.g.

wastewater treatment, desalinization, closed-loop systems), developing new products and processes that are less water-intensive and improving reputation through participation in water management initiatives such as partnerships with government, local communities and civil society.

Climate change

Over the past 200 years, land clearing and the use of fossil fuels to meet growing energy demand has contributed to an increase in greenhouse gases (GHGs) in the Earth's atmosphere. It is generally believed that this increase is leading to climate change, with multiple adverse effects on the environment. Scenarios developed by the Intergovernmental Panel on Climate Change (IPCC) predict an increase in global mean surface temperatures of between 2.0°C and 6.4°C above pre-industrial levels, by 2100. The IPCC also forecasts increased frequency and intensity of storms, floods and droughts, as well as a rise in sea levels of between 8 and 88cm between 1990 and 2100 (IPCC 2007).

Biodiversity, ecosystems and climate are closely linked. Local and global climate cycles are influenced by the way that ecosystems sequester and emit GHGs, such as carbon dioxide (CO_2), methane (CH_4) and nitrous oxide (N_2O). Furthermore, changes in land cover can alter water cycles and rainfall patterns over time and space, contributing to droughts and floods. Deforestation reduces the ability of ecosystems to sequester CO_2, while natural processes in wetlands and in agriculture (ruminant animals and rice paddies) release CH_4. N_2O emissions come from farming systems, primarily driven by manure and fertilizer use. Increased temperatures due to climate change are expected to exacerbate biodiversity loss, modify entire ecosystems and vegetation zones, and possibly heighten the prevalence of pests and diseases such as malaria, dengue fever and cholera.

The challenge of climate change is increasingly understood by business. Leading companies have audited and reported their GHG emissions for years. Others have begun incorporating carbon prices in their project and investment appraisals. A growing number of investors and entrepreneurs are developing new climate-related businesses, including carbon accounting and trading, climate mitigation and adaptation.

Habitat change

Habitat change results from both natural disturbances (such as fires) and human activities, especially agriculture. While modern agriculture has delivered massive increases in food production, contributing to food security and poverty reduction, it has also contributed to considerable damage to biodiversity and ecosystems, primarily through land conversion. Habitat fragmentation is most severe in Europe and least severe in South America. Many countries in sub-Saharan Africa are characterized by low soil productivity and thus rely on the continuous expansion of cultivated area to meet demand for food (WBCSD et al. 2006).

Habitat change can take the form of ecosystem alteration, as in the case of deforestation and urban development, as well as fragmentation. The latter may seem less harmful, but can greatly reduce ecosystem resilience and capacity to support viable populations of wildlife. Habitat change is a particular challenge for businesses that depend heavily on ecosystem services, as these may be reduced or modified as a consequence of habitat change.

Invasive species

The introduction of certain alien species into ecosystems unaccustomed to them has been a major cause of biodiversity loss, especially on islands and in freshwater habitats. Increasing travel and trade associated with globalization and human population growth have facilitated the intentional or unintentional movement of species beyond their natural habitat. Some of these alien species have become invasive, due to lack of natural pests and predators. The result is the total domination of certain ecosystems by a few alien species, often at the expense of native plants and animals.

The economic impacts of invasive alien species and the costs of preventing or controlling them are not well documented. Recent estimates suggest that total costs are in the range of millions to billions of dollars per year (Lovell and Stone 2005). For businesses that rely on native plants and animals, the spread of invasive alien species can be a major challenge. Other companies may be affected by reductions in water availability or by fouling of equipment and infrastructure, also due to the spread of alien plants and animals.

Over-exploitation of the oceans

The oceans play a key role in climate regulation, the freshwater cycle, food supply and recreation. Coastal zones cover only 8 per cent of the Earth's surface, but the benefits they provide account for more than two-fifths of the total value of all ecosystem services. Coastal zones yield about 90 per cent of marine fisheries catch, while nearly 40 per cent of the human population lives within 100 km of a coast (WBCSD et al. 2006). Pressures on coastal zones around the world are increasing for shipping, oil and gas exploration, military and security needs, recreation and aquaculture.

Over-exploitation is the most significant pressure on marine biodiversity and also on commercial fisheries. Projections based on current rates of exploitation imply that there will be no economically viable stocks of fish or invertebrate species left by 2050.

Nutrient overloading

Chemical nutrients are essential ingredients for the supply of farmed and wild products. This includes nitrogen, phosphorus, sulphur, carbon and potassium, among others. However, human activities, particularly agriculture, have significantly changed nutrient balances and natural nutrient cycles in some regions. Over the last few decades, the flow of reactive nitrogen has doubled. Over half the nitrogen-based fertilizer ever used has been applied since 1985. Phosphorus is also accumulating in many ecosystems, due to the use of mined phosphorus in agriculture and in industrial products. While sulphur emissions have been reduced in Europe and North America, they are still rising in countries like China, India and South Africa, and in southern regions of South America.

Nutrient loading (pollution) has emerged as one of the most significant drivers of change in terrestrial, freshwater and coastal ecosystems. The introduction of nutrients can have both beneficial and adverse effects, but the beneficial effects will eventually reach a plateau (for example, beyond a certain point, additional inputs do not lead to further increases in crop yield), while harmful effects will continue to increase (MA 2005d). The impacts of nutrient loading on business are not well documented but are likely to include increased water treatment costs, reduced access to key resources (e.g. freshwater fish), and reduced value of freshwater bodies for recreation and tourism.

2.2.5 Externalities and the values at stake

Underlying all the drivers or pressures described above is the fact that many biodiversity and ecosystem values remain largely invisible to economic decision makers, including government policy makers as well as business and consumers.

The lack of property rights and prices for biodiversity and many ecosystem services can suggest that their use, whether sustainable or not, incurs low or even zero cost. In fact, the costs of biodiversity loss and ecosystem degradation are all too real, but are usually incurred by people other than the user, with no basis for the former to claim compensation or redress. The lack of property rights and prices for biodiversity and ecosystem services also weakens the motivation for individuals or businesses to adopt environmentally responsible behaviour – even when the result would be a net improvement in overall economic welfare.

This gives rise to what economists call 'externalities': the negative or positive consequences of an economic activity experienced by third parties. These occur when costs are imposed by one party upon another without agreement or compensation, or alternatively when third parties enjoy benefits from an economic activity without offering any reward or recompense. The result is that market prices and private production costs fail to reflect the full value of biodiversity and ecosystem services – economic incentives to provide biodiversity benefits are weak, as are the incentives to avoid harm.

In addition, the 'public good' character of biodiversity and many ecosystem services (i.e. their inherent 'non-excludability' and 'non-rivalry') often leads businesses to see them as purely a government responsibility. Recently, however, the sheer magnitude of business risks and opportunities associated with biodiversity loss, ecosystem impacts and ecosystem service dependencies is opening new spaces for corporate leadership.

Well-known environmental externalities include impacts on ecosystems, but also directly to humans, buildings and structures (including cultural assets), and economic activities. The impacts typically relate to adverse implications associated with air emissions, discharges, spills, land-take, noise, sedimentation and waste disposal, etc. At the same time, positive externalities may arise deliberately or inadvertently (e.g. providing wildlife habitat on non-operational 'buffer' land around a major mining operation), for which companies often do not derive a direct financial return.

Monetary estimates of the economic value of environmental externalities, both positive and negative, can help inform and motivate increased conservation efforts and more sustainable use of biodiversity and ecosystem services by business and others. As this book attempts to show, such economic evaluation is increasingly feasible but not always easy because:

- There are still many unanswered questions about the role and economic value of genetic variability, species diversity and ecosystem variety. Interactions among species and the importance of complementarities or redundancy in communities of species are largely unknown. The same is true at a genetic level (see TEEB Foundations 2010, chapter 2).

- Most indicators of biodiversity and ecosystems were not developed for economic analysis or the needs of business. There is a need for better indicators that show the relationship between biodiversity and the benefits provided to people, as well as the impacts and dependence of business on biodiversity (see Chapter 3 in this book).
- Ecosystems are subject to various pressures but vary in their resilience. If pushed beyond critical thresholds, ecosystems may be transformed into less desirable states. As with economic evaluations of climate change, conventional analysis often breaks down in the face of non-marginal changes. In such cases, economic analysis gives way to ethics (see TEEB Foundations 2010).
- Cost–benefit analysis may be unreliable when key aspects are uncertain (e.g. prices, the discount rate, critical thresholds). The challenge is to identify aspects of biodiversity loss that are 'non-marginal' and the implications of the aspects for decision making. Extinction, for example, is a non-marginal event, but not all extinctions have the same implications. What matters for business is the loss of species or ecosystem services that are essential for production, or the collapse of entire ecosystems (e.g. due to invasive alien species), or the loss of consumer confidence in a product, brand or industry due to perceived adverse impacts. Some ecosystem values may not become known until far into the future, hence there is a positive 'option value' to conserving biodiversity even when current valuations appear small.
- Many ecosystem values are context-specific, not only due to the diversity of nature but also the fact that economic values reflect the number of beneficiaries and local socio-economic contexts and culture. Hence the value of a service measured in one location cannot simply be extrapolated or transferred to other sites, unless certain adjustments are made. Cultural context also influences the response to economic arguments, as societies, communities and stakeholder groups differ in their degree of acceptance of monetary valuations.

Notwithstanding such challenges, there has been considerable progress in the valuation of environmental externalities and, in particular, ecosystem services and biodiversity. Valuation studies are available for a range of ecosystem services in several biomes and ecosystems around the globe (see TEEB Foundations 2010). Such studies can give impetus to public policy making, but may also be of value for business decisions.

Making the economic value of ecosystems and biodiversity explicit can help generate support for new instruments and approaches that change the decision equation facing landowners, investors and other users of biodiversity and ecosystem services. Appropriate policy responses can take many forms, including payments for ecosystem services, reform of environmentally harmful subsidies or the introduction of resource-user charges, pollution taxes and offset requirements (see TEEB in Policy Making 2011, chapters 5–7). These and other potential policy reforms have significant implications for business.

Business can also use economic valuation to inform its own decisions. As described in this report, ecosystem valuation methods are being used by some pioneer companies to identify risk, improve operational efficiency or develop new business ventures (see Chapters 3, 4, 5 in this book).

Ecosystem valuation is also essential for understanding the true value of environmental assets and business impacts. For example, in a report prepared by Trucost for the Principles for Responsible Investment (PRI) and the United Nations Environment Programme Finance Initiative (UNEP FI), the world's 3,000 largest publicly listed companies were estimated to have caused environmental damage worth US$2.15 trillion in 2008 (see PRI and UNEP FI 2011). Of this, the major contributing costs are greenhouse gas emissions, overuse and pollution of water, and particulate air emissions. Only a proportion of these costs relate to ecosystem services (e.g. water-related impacts, as water is a provisioning service).

The Trucost analysis includes a breakdown of estimated environmental externalities for five major industrial sectors, revealing that some sectors are more exposed to potential environmental liabilities than others (Figure 2.4). This analysis also shows the relative significance of climate change in estimates of environmental impacts, perhaps mostly due to the lack of reliable monetary estimates of industrial impacts on biodiversity and ecosystems services.

Being able to assess and measure the value of ecosystems from an economic perspective provides information that can be directly integrated with conventional financial measures and unequivocally linked to the financial bottom line of business, as discussed further in Chapter 3.

2.3 Impacts and dependence on biodiversity and ecosystem services across sectors

The links between business, biodiversity and ecosystem services vary across sectors, and even within sectors. They depend on the location of the business, the source of its raw materials, in some cases the location of its customers, and/or the production technology employed.

Broadly, these links can be grouped into business impacts on biodiversity, on the one hand, and business dependence on ecosystem services, on the other. In each case

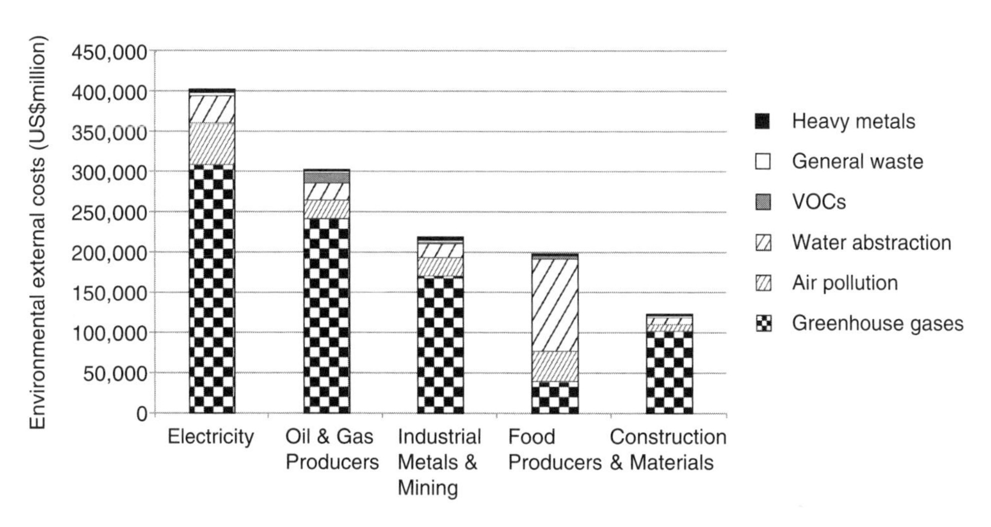

Figure 2.4 *Estimated environmental damages for five major industry sectors*

Source: PRI and UNEP FI (2011)

it is important to look beyond direct impacts and dependence, to consider the indirect links arising through business value chains. This section provides an overview and some concrete examples of business impacts and dependence on biodiversity and ecosystem services across a range of sectors.

2.3.1 Agriculture

The agriculture sector faces a growing dilemma: it needs to feed a rapidly growing and increasingly affluent global population while also conserving biodiversity and sustainably managing natural resources on an increasingly depleted planet. The need for increased food production is meanwhile constrained by poor land management and lack of means (financial and technological) to maintain, let alone enhance, productivity.

Agriculture as a sector is the biggest land manager in the world, and provides important habitats for many wild plant and animal species. At the same time, extensive land clearing for farming, and especially for large-scale intensive crop production and livestock, is one of the main drivers of biodiversity loss. Examples include the conversion of large areas of the Amazon rainforest and Brazilian savannah to soybean and cattle production, and conversion of lowland rainforest in southeast Asia to oil palm plantations. Not all recent agricultural land conversion is attributable to food production; some is also due to expanding demand for bio-fuels.

Crop and livestock genetic diversity are also declining, due to the narrowing of varieties and land races in agricultural systems. For example, more than 60 breeds of livestock are reported to have become extinct since 2000 (CBD 2010).

Agricultural productivity is heavily dependent on numerous species and ecosystem services, including soil micro-organisms, natural and domesticated pollinators and pest predators, the genetic diversity of crops and livestock, as well as freshwater supplies, climate regulation and nutrient cycling.

Insect pollinators, in particular, are estimated to provide services worth about US$189 billion per year to global agriculture (Gallai et al. 2009) through increased yields and other benefits. Loss of native bees and other pollinators is linked to loss of suitable habitat, agricultural intensification, urbanization and climate change, among other factors (UNEP 2010).

For example, a study of the climate risks to agribusiness in Ghana found that pollination services provided by a naturally occurring weevil (*Elaeidobious kamerunicus*) account for about 20 per cent of oil palm yields (Stenek et al. 2011). While the sensitivity of this insect to climate change is not well documented, a scenario leading to the total loss of the pollinator would imply a comparable reduction in yield, as producers would need to rely instead on hand pollination.

Even managed honey bees that are imported to provide pollination services have experienced steep declines (Pettis and Delaplane 2010). Grower economics, food security and biodiversity may be at risk without sustainable solutions to pollinator declines (Box 2.1).

Agriculture is a major contributor to climate change, responsible for around 14 per cent of global GHG emissions in 2000, from sources including fertilizers, livestock, wetland rice cultivation, manure management, burning of savannah and agricultural residues, and ploughing (IPCC 2007). Conversion of forests to agriculture, primarily in developing countries (and particularly in tropical Asia), accounts for a comparable share of greenhouse gas emissions (Hanson et al. 2008; Werf et al. 2009).

Box 2.1 Operation Pollinator: Investing in natural capital for sustainable agriculture

Syngenta, a supplier of seeds and crop-protection products to the agriculture sector, is supporting a novel conservation programme, which aims to enhance farm productivity by safe-guarding a key ecosystem service essential to agriculture – namely pollination of crops by insects. From field vegetables in Italy to melons in France or blueberries in the United States, the importance of insect pollination for agricultural productivity is unequivocal.

In 2009, Syngenta launched Operation Pollinator, an initiative currently involving 13 countries in the EU, as well as the USA. The initiative aims to restore healthy populations of native (wild) pollinators in agricultural landscapes, by creating suitable habitats on or near farmland. By establishing and managing floral plant margins around crops, plant diversity is increased and populations of native pollinators are boosted, generating significant environmental benefits and potential increased farm profits.

The benefits of Operation Pollinator are clearly apparent in the state of Michigan, home to the largest blueberry industry in the USA. The economic value of insect pollinators in Michigan is substantial: the fruit and vegetable sector is worth around US$800 million annually and includes several crops that are highly dependent on pollination to sustain marketable yields. In Michigan, the blueberry crop is 90% reliant on pollination by bees, which ensure high yields and a crop worth over US$100 million annually (USDA NASS 2008; AgMRC 2010). With continuing losses of managed honey bee colonies (vanEngelsdorp et al. 2011), new crop pollination strategies are needed to sustain agricultural production.

Operation Pollinator advises and trains growers interested in converting marginal agricultural land to native pollinator habitat. In coordination with federal conservation programmes, growers are assisted to make simple changes in farm operations that are compatible with current practices and existing soil and water conservation goals. Enhanced native pollinator populations are part of a diversified strategy for maintaining crop yields and improving fruit quality for pollination-dependent crops. When managed bees are in short supply, or if beehive rental costs increase, abundant native bee populations can provide supplementary pollination services to farmers.

The future of agriculture ultimately depends on protecting the environment and enhancing the livelihoods of growers, through the development of sustainable agricultural systems. Modern agriculture increasingly recognizes the commercial benefits of farm management practices that increase the diversity of beneficial insect species, while also conserving other resources (e.g. soil and water). If this can be done in a way that enhances long-term productivity, not only Syngenta but also the growers it supplies and society as a whole will reap the rewards.

For more information see: www.operationpollinator.com.

Blueberry production in Michigan

- In 2006 there were 575 blueberry farms across the state, with 18,500 acres under production.
- The average yield of blueberries in 2009 was 5,350 pounds/acre.
- In 2007–2008 fresh blueberries retailed above US$2.00/pound.
- Pollination by managed bees costs Michigan blueberry growers about US$80/acre/year (based on 2 hives/acre).

Source: Peters et al. (2010) for TEEB

Other agricultural impacts arise from the introduction of invasive alien species, soil and water pollution due to inefficient use of pesticides and fertilizers, erosion and/or sedimentation of downstream ecosystems, including wetlands and even offshore coral reefs (Box 2.2). Overall, about 85 per cent of agricultural land is considered to be degraded due to erosion, salinization, soil compression, nutrient depletion, biological degradation or pollution, while each year 12 million hectares are lost to desertification (Hanson et al. 2008).

Although agriculture is often characterized by adverse impacts on biodiversity, there is growing interest in finding ways to combine intensive arable production with biodiversity conservation. An example is provided by the BASF Biodiversity Project on Rawcliffe Bridge Farm in the UK (BASF n.d.). Since the project began in 2002, indicators of biodiversity have improved, including notably an increase in the number of recorded bird species from 80 to 103, as well as the first recording of 160 species of moths, 21 species of butterflies and 2 species of bats. Improvements in agronomic practices include reduced application of summer insecticides, direct cutting of winter oilseed rape instead of swathing, and delaying the dredging and cutting of farm dykes (ditches). Additional examples of efforts to integrate biodiversity in agricultural production systems are described in Chapter 5 of this book.

2.3.2 Forestry

Forest ecosystems are important habitats for many wild plant and animal species, and the forestry industry depends on numerous ecosystem services, including fresh water supply, climate regulation and nutrient cycling.

Sustainable forest management has a large role to play in biodiversity conservation and mitigating climate change. On the latter point, not only do forests and wood products act as carbon sinks, but the main industrial outputs of the forest products industry – timber and pulp – are renewable if managed sustainably. In addition, compared with other common building materials such as cement, steel and aluminium, wood-based building materials require less energy for production, have higher thermal efficiency and can be reused, recycled or used as biomass for energy.

While forest loss is primarily driven by expanding agriculture, unsustainable commercial logging activities are nevertheless a significant contributor to forest and biodiversity loss around the world (Box 2.3). In south and southeast Asia and the Pacific, for example, unsustainable commercial wood extraction is thought to account for about 25 per cent of deforestation (Mardas et al. 2009).

Such activities are often driven by illegal logging, which particularly occurs in parts of Asia, Russia, central and eastern Europe, and central Africa. Illegal activities of special concern for forest biodiversity include harvesting timber without authority in protected areas, harvesting without or in excess of concession permit limits, failing to report harvesting activity, and violating international trade agreements such as the Convention on International Trade in Endangered Species of Wild Fauna and Flora (CITES). Such activities have been estimated to represent 5–10 per cent of global industrial roundwood production (Seneca Creek Associates 2004).

Both legal and illegal logging activities can also indirectly affect biodiversity through the construction of logging roads (habitat fragmentation), which can facilitate small-scale mining, hunting, illegal logging, fishing and settlement within previously untouched forests.

Box 2.2 Cotton production and desiccation of the Aral Sea

A striking example of how overuse of fresh water can destroy an entire ecosystem is the desiccation of the Aral Sea due to unsustainable agriculture. Located between Kazakhstan and Uzbekistan in Central Asia (formerly part of the Soviet Union), the Aral Sea was the world's fourth largest inland sea in 1960, providing a wealth of products and services. Over the period 1960–1990, however, the development of irrigated agriculture around the Aral Sea increased from about 4.5 million ha to just over 7 million ha, while the surface area of the Aral Sea declined from almost 70km^2 to under 40km^2. By 2007, the Aral Sea had shrunk to about 10% of its original size, mainly due to water abstraction from two major tributaries, the Amu Darya and Syr Darya rivers, for irrigated cotton production (Micklin and Aladin 2008).

The expansion of cotton production led to increased use and runoff of pesticides and fertilizers, resulting in pollution of both surface and groundwater. Water abstraction for irrigation reduced the flow of water downstream, drying out lakes and wetlands, while also increasing salinity. As a result, both the Amu Darya delta in Uzbekistan and the Syr Darya delta in Kazakhstan suffered substantial damages. Between 1960 and 1990, some 95% of the wetlands surrounding the Amu Darya delta disappeared, while more than 50 delta lakes, covering 60,000ha, completely dried up (FAO 1998). Some 100,000ha of Tugai forests, which covered the Amu Darya delta in 1950, were reduced to less than 30,000ha by 1999 (Severskiy et al. 2005). The lakes of the Syr Darya delta were also severely affected, shrinking from about 500km^2 to 40km^2 between 1960 and 1980 (Micklin 1992).

A study in 1990 suggested that environmental damages due to unsustainable agriculture and poor irrigation practices around the Aral Sea amounted to at least US$1.4 billion, based on an estimate of the cost of measures to address at least some impacts (Glazovsky 1990). A follow-up study estimated the cost to improve sanitary, hygienic and medical services in the region, create alternative employment, and shift the economy onto a more sustainable path was over US$3.5 billion (Glazovsky 1995). Yet another, more recent study focuses on the costs of redirecting water from the Volga, Ob and Irtysh rivers to restore the Aral Sea to its former size, estimated at over US$30 billion (Temirov 2003).

These studies are instructive, but they do not directly consider the value of ecosystem services lost due to desiccation of the Aral Sea. For TEEB, a meta-analytic value function was used to estimate the decline in ecosystem services resulting from the loss of 522,500ha of wetlands over the period 1960–1990. The results suggest annual losses of about US$100 million (Brander 2010). This is a conservative estimate, including only a portion of the ecosystem services lost, and may not reflect the specific characteristics of the Aral Sea basin. Nevertheless, even this conservative assessment suggests that ecosystem impacts due to cotton production in the Aral Sea basin are economically significant. In addition, unsustainable irrigation practices have resulted in significant on-farm losses. These include water-logging, increased soil salinity and reduced crop yield, with lost crop production alone valued at US$1.4 billion per year, or about one third of the value of potential crop production (Glazovsky 1990).

Note that, in the case of Uzbekistan, over 70% of cotton production (or around 800,000 tonnes per year) is exported, making it the world's second largest cotton exporter (Environmental Justice Foundation 2005). The largest consumer of Uzbek cotton is the European Union, which accounts for 29% of Uzbek cotton exports, or about US$350 million per year (Environmental Justice Foundation 2005). What is not clear is how this trade would be affected if the ecosystem impacts of cotton production were included in the price of Uzbek cotton.

Source: See Annex 2.1, Case Study 1

Box 2.3 The construction industry and deforestation in China

Over the period 1949–1981, China significantly depleted its natural forests in order to meet demand for timber for construction and other uses. A cumulative area of 75 million hectares was harvested, of which 92% were natural forests (Song and Zhang 2010). This deforestation resulted in: (1) reduced per-hectare timber stocks, (2) an age structure tilted towards younger forest stands, (3) changed species composition, (4) reduced rates of regeneration, and (5) lower growth and yields of forest plantations (Yin 1998). Furthermore, ecosystem services provided by forests, such as watershed protection and soil conservation, were severely compromised. The conversion and degradation of natural forests reached a tipping point in 1997, when severe droughts led to the lower reaches of the Yellow River drying up for 267 days, threatening industrial, agricultural and residential water users in the northern plains (Xu and Cao 2002). The next year, flash flooding occurred in almost all major river basins in China, devastating large areas, causing damages estimated at 248 billion yuan (approximately US$30 billion in 1998), a loss of over 4,000 human lives and displacement of millions of people (Sun et al. 2002).

Following these disasters, the Chinese authorities identified deforestation and farming on steep slopes as the primary causes of both the Yellow River drought in 1997 and the flooding of the Yangtze River basin in 1998 (Larsen 2002). In particular, logging in areas around major river systems was identified as a cause of increased runoff, soil erosion and siltation of waterways (Lang 2002). In response, the government banned logging outright in 17 provinces, under the 1998 Natural Forest Conservation Program (NFCP). Timber harvests from natural forests in China fell from 32 million m^3 in 1997 to 12 million m^3 by 2003. The restrictions on logging were also reflected in 20–30% higher prices for timber in the Beijing wood market.

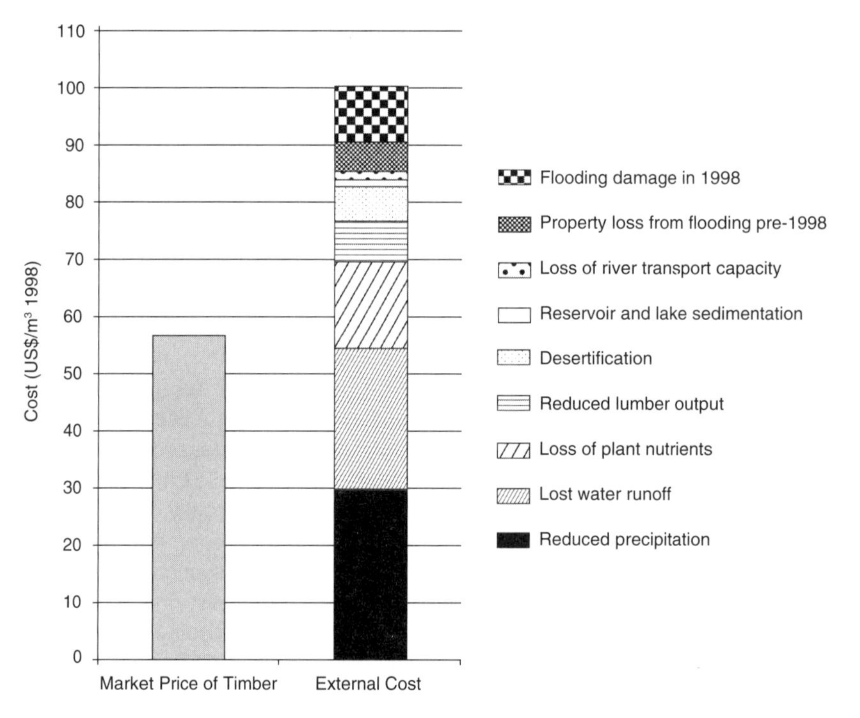

Comparing market prices with ecosystem externalities: The case of timber in China

Source: See Annex 2.1, Case Study 2

The economic costs of deforestation in China were examined by another study, which considered the various ecosystem services lost, such as climate regulation, provision of timber and food, water regulation, erosion control, flood prevention and nutrient cycling (Wang Hongchang 1997). Based on this study, the value of ecosystem services lost due to logging for the Chinese construction and materials industries was estimated at US$12.2 billion annually (McVittie 2010). If we then compare the value of ecosystem service losses per cubic metre of timber with the prevailing market price of timber, it is evident that the full costs of logging were not reflected in the prices paid for timber by the construction and materials sectors (see chart). Put another way, if the external costs of ecosystem degradation due to unsustainable logging and deforestation were reflected in the price of timber, we would expect to see a significant price hike!

The forestry industry, especially the pulp and paper sector, is relatively energy intensive. However, wood-based biomass is often used as a fuel that, if sourced from a sustainably managed forest, is carbon-neutral, or near carbon-neutral. Overall, production processes in the forest products industry make a relatively small contribution to climate change, being responsible for about 1.6 per cent of global CO_2 emissions (WBCSD 2007).

2.3.3 Mining and quarrying

With the exception of supplies of fresh water for mineral processing, the mining industry is not directly dependent on biodiversity or ecosystem services. However, it has a number of significant direct and indirect impacts on BES that, if ignored, can create major risks to mining operations.

One of the main direct impacts comes from surface mining, whereby overlying habitats and geological features are removed during the extraction of minerals. Other disturbances to plants and animals during the quarrying process include noise, dust, pollution, and removal and storage of waste (tailings).

While the process of quarrying itself is mainly associated with negative impacts, a growing number of companies are beginning to use biodiversity offsets to compensate for residual impacts that cannot be mitigated on site, while also investing in ecological restoration and rehabilitation on former mine sites. In some cases, such activities can deliver significant biodiversity value (Box 2.4).

The indirect impacts of mining on biodiversity and ecosystems include pollution and water used during refining and smelting processes. Mining operations can also have significant indirect impacts on biodiversity-sensitive areas by developing roads that provide access to areas that were previously underdeveloped and inaccessible, leading to immigration and accelerated conversion of habitat.

2.3.4 Oil and gas

The oil and gas sector delivers a range of products to end-users, primarily energy for transport and power, but also petrochemical-derived products such as plastics. Most stages of oil and gas production and distribution depend little on biodiversity and ecosystems services as direct inputs, although freshwater supply is important in some cases.

Box 2.4 Holcim and the value of wetland restoration

Aggregate Industries UK, a subsidiary of Holcim, operates the Ripon City Quarry in the Yorkshire Dales in England, mining sand and gravel since 1964. As part of legal requirements and its commitment to sustainable development, where relevant the company systematically restores ecosystems as part of its extraction operations. Current UK legislation requires the company to provide a rehabilitation plan when applying for a new or extension of permit. In the Ripon City Quarry, in support of a request to extend the existing quarry over land currently used for agriculture, the company proposed to create a mix of wetlands for wildlife habitat and an artificial lake for recreation, following mineral extraction. However, in the process of stakeholder discussions, disagreement developed within the local conservation community regarding the most appropriate restoration options. Ecosystem valuation was considered a useful tool to help identify the restoration option with the greatest benefits to local communities and the region.

A valuation study, carried out with the support of IUCN, aimed to quantify in monetary terms the impacts of quarrying and restoration operations on biodiversity and ecosystem services. The study considered three sites proposed for the extension of the quarry, together covering roughly 38 hectares. While all sites are currently used for agriculture, only the least flood-prone site was to be restored to agricultural use. A second site was to be restored to wetland and managed for biodiversity. An artificial lake managed for recreation (sailing) was planned for the final site, in part due to the depth of planned extraction activities and the lack of available soil/overburden for restoration. The analysis included all significant costs and benefits within the boundaries of the extension sites. Costs included both investment and recurrent costs associated with restoration and the opportunity cost of the land (for agricultural production). Benefits generated within the restored extension sites were included, as well as off-site benefits to the local community and further afield. The legal requirement is for a 50-year aftercare programme, hence the study took 50 years to be the planning horizon. Moreover, due to limited time and resources, a 'benefits transfer' approach was adopted for this case study. This means that unit estimates of the value of ecosystem services (e.g. provision of wildlife habitat in predominantly agricultural landscapes, flood control, carbon storage, and recreational opportunities associated with artificial lakes) were taken from detailed studies of similar sites in the UK and adjusted to local conditions.

The main benefits valued were wildlife habitat provided by the wetland and recreation and flood control benefits provided by the planned artificial lake. The value of carbon sequestered in newly established wetlands was also valued. Each site was valued on the basis of one or two dominant ecosystem services, in order to avoid double counting when using willingness to pay (WTP) estimates for wetlands, which typically include values for a range of ecosystem services.

The study showed that the value of biodiversity benefits expected to be generated by the proposed wetlands (£1.4 million), the recreational benefits of the artificial lake (£350,000) and the increased flood storage capacity of the overall area (£224,000) would, after deducting restoration and opportunity costs, deliver net benefits to the local community of about £1.1 million, in present value terms, using a 50-year time horizon and a 3 per cent discount rate (see chart). The value of carbon sequestration in these wetlands was found to be relatively small, while the marginal benefits associated with wetlands far exceeded the current benefits derived from agricultural production. The study further showed that the costs of ecosystem restoration and aftercare are low, compared with both the economic benefits of wetland restoration and the financial returns from sand and gravel extraction.

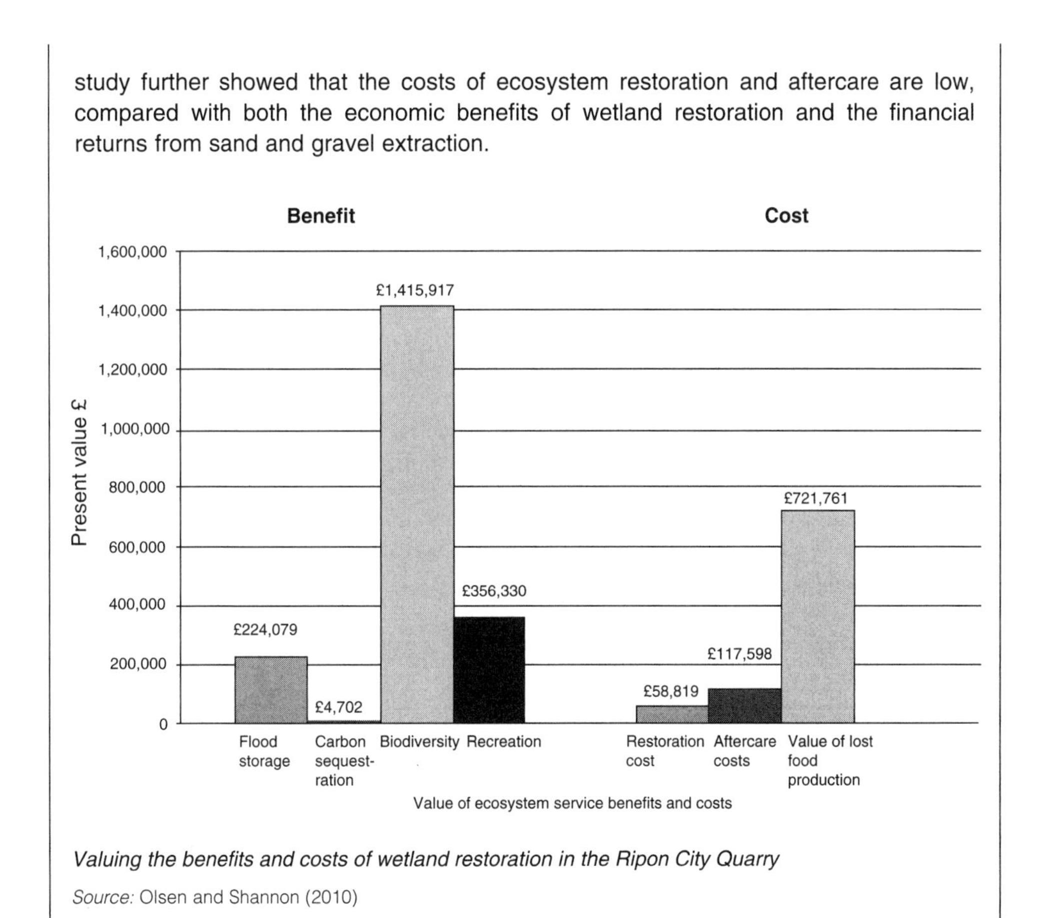

Valuing the benefits and costs of wetland restoration in the Ripon City Quarry

Source: Olsen and Shannon (2010)

In terms of its direct impacts on biodiversity, the industry can be split between upstream (exploration and production) and downstream (refining, marketing and distribution) activities. The most visible potential impacts on biodiversity are from upstream activities (seismic studies, drilling, construction and production), typically in relatively undeveloped areas. While significant progress has been made in much of the sector to reduce these impacts through improved operating practices, increasing exploration and production in environmentally sensitive areas (e.g. deep water, the Arctic and boreal forests) risks direct impacts on biodiversity and ecosystems (Box 2.5). Secondary impacts arising from human activities linked to or attracted by oil and gas projects are also a risk. At a global level, emissions of greenhouse gases from the consumption of oil and gas products is linked to increasing stress upon biodiversity.

2.3.5 Cosmetics and personal care

The cosmetics sector relies on biodiversity for many natural ingredients. For instance, cosmetic companies such as Estée Lauder and L'Oréal use a succulent plant listed in

Box 2.5 Valuing the ecological impacts of the Deepwater Horizon oil spill

The Deepwater Horizon oil spill in the Gulf of Mexico, also known as the Macondo Incident, was the largest accidental marine oil spill in history. The immediate cause of the spill was a blowout and explosion on 20 April 2010, which killed 11 crewmen and destroyed the Deepwater Horizon offshore oil platform. After a series of failed efforts the leak was finally plugged in July 2010. While the total amount of oil spilled was the subject of debate, the current consensus is that roughly five million barrels of oil were released, of which some four million barrels ended up in the waters of the Gulf of Mexico following short-term clean-up efforts (Cleveland 2010).

The Deepwater Horizon platform was owned by Transocean Ltd but under contract with BP as the principal developer. BP was held responsible by the US government for clean-up costs and other damages; the resulting litigation is expected to take years. The eventual costs of the spill to BP and its shareholders, to communities in affected areas, and to society as a whole are not known. Nevertheless, it is already clear that the spill had major economic consequences.

Looking first at BP itself, the company's group income statement for the second quarter of 2010 showed a pre-tax charge of US$32.2 billion related to the oil spill. This includes US$2.9 billion of clean-up costs incurred through 30 June 2010. Following standard accounting rules, BP treated all charges relating to the incident as non-operating items to be deducted from taxable income. This includes a US$20-billion escrow account set up to fund approved damage claims. In addition, BP committed up to US$500 million for a 10-year research programme on the impact of (a) the Gulf of Mexico oil spill and (b) its associated response on the marine and shoreline ecosystems. The company also agreed to fund the US$360 million cost of installing six berms in the Louisiana barrier islands project. It remains unclear how much the company and its shareholders will ultimately have to pay.

More broadly, the full ecological and economic impacts of the oil spill are difficult to quantify. One preliminary assessment suggests that the spill adversely affected some 20 categories of ecosystem services in and around the Gulf of Mexico, including temporary closure of the US$2.5 billion per year Louisiana commercial fishery (Costanza et al. 2010). The same study goes on to consider total losses, building on earlier calculations of the total value of ecosystem services for the Mississippi River Delta. These were estimated at US$12–47 billion per year (Batker et al. 2010). Taking this range of values as their starting point, assuming further that the Delta will be the most affected region and that there will be a 10–50 per cent reduction in the value of ecosystem services as a result of the oil spill, Costanza et al. estimate the total value of ecosystem service losses at between US$1.2 billion and US$23.5 billion per year until ecological recovery occurs, or US$34–670 billion in present value terms (using a 3.5 per cent discount rate; see Costanza et al. 2010 for details).

In practice, governments and the courts can find it difficult to measure losses of non-market ecosystem services reliably (Boyd 2010), and both the methods and results of such studies are often contested (Chevassus-au-Louis et al. 2009). As an alternative, many agencies assess environmental damages using a resource replacement cost approach, in which the goal is to replace lost economic and ecological wealth through restoration. In other words, by soliciting restoration bids and using those bids to guide damage negotiations, agencies can avoid the difficulty of measuring non-market wealth.

Using restoration costs to measure environmental damages is no panacea and can result in significant over- or under-deterrence, depending on the relationship of restoration costs to the social cost of physical damages (Boyd 2010). Until methods for measuring ecosystem services are more mature, however, plaintiffs, trustees and courts are more likely to resolve damage claims through legal and political bargaining rather than technical calculation.

Whatever the outcome of damage claims arising from the Deepwater Horizon oil spill, we may expect more stringent regulation of industrial activities in the USA and elsewhere, particularly in relation to deep water oil and gas exploration. As noted by BP itself (2010, p. 33):

> *significant uncertainties over the extent and timing of costs and liabilities relating to the incident and the changes in the regulatory and operating environment that may result from the incident have increased the risks to which the group is exposed. These uncertainties are likely to continue for a significant period. These risks have had and are expected to have a material adverse impact on the group's business, competitive position, cash flows, prospects, liquidity, shareholder returns and/or implementation of its strategic agenda.*

Such an acknowledgement highlights the case for integrating biodiversity into business risk management and information systems.

Source: Houdet (2010) for TEEB

CITES Appendix II (*Euphorbia antisyphilitica*) whose distinctive properties make it an essential raw material in a wide array of cosmetics (especially lipsticks), inks, dyes, adhesives, coatings, emulsions, polishes, pharmaceutical products and gum base. Similarly, the essential oil of holywood (*Bulnesia sarmientoi*) and Brazilian rosewood (*Aniba rosaeodora*) trees is used in the perfume industry because of its mild and pleasant fragrance.

As the variety of species and habitats continues to decline, both the quality and quantity of these natural ingredients may be jeopardized. In its activity report for 2008, Colipa, the European Cosmetics Association, identified respecting the scarcity of natural resources, reducing biodiversity damage and developing resource efficient product life cycles as among the primary challenges for the sector (Colipa 2008).

According to the Union for Ethical BioTrade, the top 20 cosmetics and personal care companies communicate the most about biodiversity (Union for Ethical BioTrade 2010). This number is likely to increase as a growing number of companies in this sector take steps to integrate the principles and practices of sustainable use in their supply chains.

Brazilian cosmetics company Natura, for example, adopted the sustainable use of biodiversity as the main driver of innovation. Natura developed vegetable renewable alternatives to petrochemical raw materials, reducing the company's carbon footprint, and created an entire product line (Ekos) based on the sustainable use of biodiversity. Ekos has since grown to account for a substantial share of company sales (Natura 2008). Another company, L'Oréal, has developed approaches aimed at ensuring sustainable sourcing practices for plant-based ingredients (L'Oréal 2009).

A renewed focus on natural ingredients is apparent throughout the food and cosmetics sectors. Consumer concerns about health and adoption of more 'wholesome' lifestyles are important market drivers, as affluent consumers are keen to buy products that enhance (or are perceived to enhance) their well-being.

2.3.6 Pharmaceuticals

The pharmaceutical sector develops, manufactures and sells products intended to prevent, cure or relieve human and animal disease. Between 1981 and 2006, almost half of all cancer drugs and about one-third of all small-molecule new chemical entities for all disease categories were either natural products or directly derived therefrom (Newman and Cragg 2007). Examples include the common painkiller aspirin, derived from willow bark, and paclitaxel, an anti-cancer drug derived from yew trees (Sampath 2005). Although the industry has moved away from reliance on natural product discovery in recent years, it nonetheless continues to play a part in drug development and manufacturing. According to one estimate, at current rates of plant and animal extinction, we may be losing one major drug every two years (UNEP-WCMC 2002).

Potential impacts of the pharmaceutical sector on biodiversity and ecosystem services include over-exploitation of natural sources of active ingredients (where domestication or artificial synthesis is not possible), soil and water contamination from industrial production and product disposal, and impacts arising through the supply chain (e.g. soya, fish, palm oil).

2.3.7 Water supply and sanitation

The water sector is highly dependent on ecosystems for sustainable and cost-effective operations. The quantity and quality of water depends on functioning aquatic ecosystems, including lakes, rivers, streams and wetlands, as well as local bio-physical processes and land-use practices (Box 2.6). Ecosystem services important to water utilities include:

- protection of water quality and quantity through the water cycle and hydrological processes throughout a catchment area;
- riparian vegetation filtering water, removing impurities and reducing erosion;
- catchment rehabilitation processes following natural events such as flood and fire;
- flood protection;
- assimilative capacity of large waterways and the ocean for wastewater discharge and treatment;
- microbiological purification of waste water; and
- other ecosystem services associated with intact watershed catchments (e.g. carbon sequestration, pollination, bio-banking and biodiversity offset value, recreational and cultural values).

2.3.8 Fisheries

According to the Food and Agriculture Organization of the United Nations (FAO), the world's capture (non-aquaculture) fisheries produced 92 million tonnes of fish

Box 2.6 How water users depend on ecosystem services: Examples from Australia

Murray Darling River Basin

The Murray Darling River Basin in Australia comprises just 6 per cent of the total surface water runoff in Australia, but supports 75 per cent of the nation's irrigated agriculture, valued at almost US$9 billion per year. Prior to a cap on water diversions, introduced in 1995, the growth in irrigation development had led to severe over-allocation of water resources, resulting in the degradation of river and wetland ecosystems, and reduced security of supply. The overuse of water has been exacerbated by droughts and climate change. In response, the Commonwealth government invested over US$11 billion in programmes to improve the efficiency of irrigation systems and to purchase water entitlements from irrigators. In New South Wales, for example, the State and Commonwealth governments provided US$95 million in structural adjustment payments to irrigators, as part of a programme to reduce water allocations in groundwater systems to more sustainable levels (Sutherland 2007).

Melbourne

The City of Melbourne draws a significant proportion of its water supply from watersheds in protected forests. These forests provide natural filtration of the water flowing through the catchments. If these watersheds were logged or if the land was converted to agricultural or urban development, it is estimated that Melbourne would need to build new water treatment facilities at a cost of about US$1 billion, with additional operating costs running into hundreds of millions of dollars each year (Young 2003).

in 2006, of which 81.9 million were harvested at sea. In that year, the total value of the marine and freshwater catch at the first point of sale was estimated at US$91.2 billion. However, the FAO also estimates that some 52 per cent of commercial fish stocks or species groups are fully exploited, 19 per cent are over-exploited and 9 per cent are depleted or recovering from depletion. The maximum wild capture fishing potential from the world's oceans has probably already been reached and a more sustainable approach to fisheries management is urgently required (FAO 2009).

Unilever, an international producer of food, home care and personal care products, is one of many companies directly affected by overfishing. Cod was the main fish species used in the company's premium frozen food products. In the 1990s, cod stocks declined precipitously and collapsed altogether in the western North Atlantic, due to mismanagement and over-exploitation. As a result, cod prices have increased substantially since the early 1990s, and by over 50 per cent between 1996 and 2000. These dramatic price increases are thought to have reduced the profit margins on Unilever's cod-related products by about 30 per cent (F&C Investments 2004).

2.3.9 Tourism

The global tourism industry generated about US$5.7 trillion of value added in 2010 (over 9 per cent of global GDP) and employs around 235 million people directly or indirectly (WTTC 2010). Many tourism businesses are fully or partially dependent on biodiversity and ecosystem services, relating to ecotourism, beach holidays, skiing, visiting national parks, etc. Whale watching alone was estimated to generate US$2.1 billion per year in 2008, with over 13 million people undertaking the activity in 119 countries (IFAW 2009).

On the other hand, tourism developments and associated activities can cause considerable damage to biodiversity and ecosystems through land conversion, water use, sewage and solid waste. In addition, tourism businesses that depend on ecosystem services are highly susceptible to external impacts on biodiversity. The situation is particularly acute for tourism businesses that depend on coral reefs, due to the pressure posed by climate change (Wilkinson 2008). For example, coral reefs in the Caribbean are thought to have receded by about 80 per cent over the past three decades, resulting in a reduction in revenues from dive tourism (which account for almost 20 per cent of total tourism receipts) estimated at about US$300 million per year (UNEP 2008).

One eco-resort company, Chumbe Island Coral Park Ltd, has identified both the opportunities and risks associated with dependence on fragile coral reefs. The company has invested over US$1.2 million to establish a marine park in order to protect the coral reefs surrounding Chumbe Island, just off Zanzibar in Tanzania. The company manages the park with 11 park rangers and restricts guest numbers to a maximum of 16 at any one time. The business currently generates over US$500,000 per year in revenues and employs 43 staff in total, all but two of whom are locals (Riedmiller 2010).

In a similar vein, public–private partnerships are increasingly being developed between tourism companies, governments and NGOs to protect ecosystems in order to support businesses and livelihoods. For example, ecotourism ventures in Rwanda and neighbouring countries, oriented around the viewing of endangered mountain gorillas, can generate about US$500 per tourist visit. The revenues obtained from ecotourism are an important source of support for local communities in the area and for government agencies involved in gorilla conservation.

2.3.10 Transport

Transport within business comprises specialist transport and logistics companies, as well as distribution, logistics and transport departments within larger companies. Covering air, road, rail and sea transport, this sector is characterized more by its impacts on biodiversity than by its dependence on ecosystem services. Typical risks to biodiversity relate to transport infrastructure (e.g. land take and pollution associated with developing and operating roads, ports and depots), incidents such as ship groundings on corals and oil spills, and operational externalities such as emissions of carbon, NO_x, SO_x and particulates.

Recent work by ERM and EcoConsult, relating to relocating and expanding port, ferry and container terminals in Jordan, identified potential compensation activities for impacts on coral reefs, potentially amounting to millions of US dollars, depending on the size of impact. Proposals were also developed to transplant coral and create artificial reefs to offset part of the damages.

Oil spills and ship groundings on reefs can lead to multimillion dollar compensation and remediation claims (IOPC Fund 2009; Spurgeon 2006). With regard to day-to-day operational impacts, the impacts of road transport on air pollution are under increasing scrutiny, as reflected for instance in the proposed EU Eurovignette Directive targeting Heavy Goods Vehicle emissions (Grangeon and Cousin 2009).

2.3.11 Manufacturing

The impacts of the manufacturing industry on biodiversity are both direct and indirect, reflecting the footprint of facilities and the pollution arising from production processes, as well as the impacts of suppliers of raw materials or semi-finished goods. These linkages are often complex and sector specific, as seen in the case of office equipment manufacturer Ricoh and illustrated by the 'biodiversity relevance map', which was developed for Ricoh by the Japan Business Initiative for Conservation and Sustainable Use of Biodiversity (JBIB) (Figure 2.5).

Figure 2.5 also highlights some of the difficulties of attempting to measure the links between manufacturing, biodiversity and ecosystem services. In the case of Ricoh, the main impacts on BES arise through the procurement of raw materials (e.g. paper pulp and metals), as well as in the manufacturing process itself (particularly with regard to water resources).

2.3.12 Finance

The impacts of the financial services industry on biodiversity are indirect but nevertheless can be very significant. While financial institutions are not directly dependent on ecosystem services, they are exposed to BES risks through the loans, investments and insurance cover they provide to companies and projects.

The banking sector uses at least four strategies to manage biodiversity risks: (1) 'red-lining' investments in areas of high biodiversity; (2) developing sector guidelines for environmentally sensitive sectors; (3) refraining from financing sectors in which a bank lacks specialist knowledge; and (4) working together with borrowers to improve their environmental performance and mitigate harm through an engagement policy (Coulson 2009).

A review of 50 large banks revealed that 32 per cent of them have developed sector specific guidelines for clients and projects in the forestry sector (Mulder and Koellner 2011). For example, JPMorgan Chase sets deadlines for verifying the legal origin of wood sourced by clients from countries with a reputation for illegal logging (FSC 2005).

Similarly, HSBC has developed guidelines for its project finance activities in the fresh water infrastructure sector (HSBC 2005). These limit HSBC investments in projects located in, or impacting on, critical natural habitats, sites inscribed on the Ramsar list of internationally important wetlands, UNESCO World Heritage Sites, and dam projects that do not conform to the framework of the World Commission on Dams.

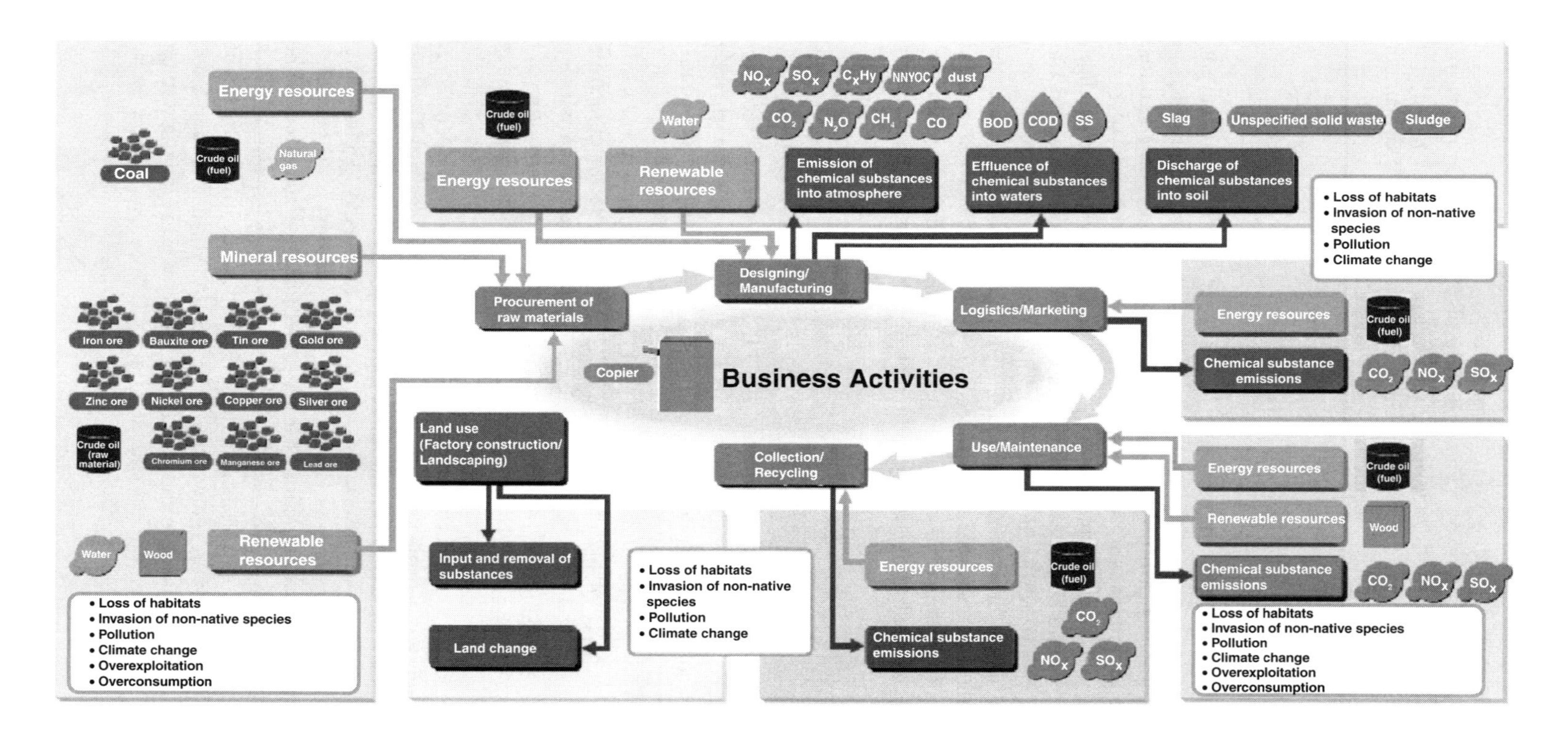

Figure 2.5 *Ricoh's map of corporate activities and biodiversity (recycled multifunctional digital copiers)*

Source: Ricoh 2009

More generally, the World Bank's International Finance Corporation (IFC) has sought to reflect the value of biodiversity and ecosystem services in a revision of the bank's influential Performance Standard 6, focusing on Biodiversity Conservation and Sustainable Management of Living Natural Resources. Thus the revised standard includes as a specific objective the maintenance of ecosystem service benefits.

2.4 Biodiversity and ecosystem risks and opportunities for business

As outlined above, companies face several risks related to biodiversity and ecosystem services (Hanson et al. 2008). These may be summarized as:

- operational;
- regulatory;
- teputational;
- market or product; and
- financial.

At the same time, it is clear that biodiversity and ecosystem services present new business opportunities, such as:

- New technologies and products – which will serve as substitutes, reduce degradation, restore ecosystems or increase the efficiency of ecosystem service use.
- New markets – such as water-quality trading, certified sustainable products, wetland banking and threatened species banking.
- New businesses – such as ecosystem restoration and environmental asset finance or brokerage.
- New revenue streams – for assets currently unrealized, such as wetlands and forests, but for which new markets or payments for ecosystem services could emerge.

All of these are discussed in general terms below, with examples from several companies. Further detail on the tools and approaches available for business to reduce their BES risks and seize BES opportunities is provided in Chapters 4 and 5, respectively.

2.4.1 Operational

- Risks – increased scarcity and cost of raw materials such as fresh water, disruptions to business operations caused by natural hazards and higher insurance costs for disasters such as flooding.
- Opportunities – benefits from increasing water-use efficiency or building an on-site wetland to eliminate the need for new water treatment infrastructure.

In 2001, Anheuser-Busch, the world's largest brewer of beer, experienced unexpected water shortages that affected its supply chain. A temporary drought in the US Pacific Northwest increased the price and reduced the availability of key inputs to Anheuser-Busch's brewery operations, namely barley and aluminium. This example highlights the need for business to take a comprehensive view of its dependence on ecosystem services (Global Environmental Management Initiative 2002a).

Wetlands are known for their ability to clean water, absorb waste and breakdown some pollutants. DuPont built a wetland to treat wastewater from a plant in Victoria, Texas, after local residents started expressing concerns about the deep well injection process the company had previously been using. After recovering over 250,000 pounds of material formerly lost to wastewater streams each day, the waste water is treated in an on-site biological facility before being released to the constructed wetlands for further natural 'polishing' prior to its eventual return to the Guadalupe River. Over 2.4 million gallons of recovered water are returned to the Guadalupe River each day (Global Environmental Management Initiative 2002b).

2.4.2 Regulatory and legal

- Risk – emergence of new fines, new user fees, government regulations, or lawsuits by stakeholders.
- Opportunity – engaging governments to develop policies and incentives to protect or restore ecosystems that provide services a company needs.

In 2004, the UK government denied Associated British Ports planning permission for a port expansion at Dibden, due to its potential encroachment on nearby coastal ecosystems that were valued for their biodiversity and associated cultural services. As a result, Associated British Ports had to write off £45 million that it had already spent on the proposal and the company's share price dropped by 12 per cent in the week immediately following the permit denial (F&C Investments 2004).

The International Maritime Organization issued regulations that took effect in 2009 in order to prevent the transport of invasive alien species via ship ballast water. Aquatic species transported from one ecosystem to another by ship can have devastating effects on marine life and on local economies. To help ship owners meet the new requirements, Alfa Laval developed and launched PureBallast, a first-to-market ballast water treatment system that removes unwanted marine organisms without additives or chemicals (Alfa Laval 2010).

Unregulated wildlife trade can seriously affect species populations, especially those that are already vulnerable as a result of other factors, such as habitat change. In 1973, governments responded to this concern by regulating international trade in wild fauna and flora through CITES, which aims to ensure that trade in products and derivatives of wild fauna and flora is sustainable, legal and traceable.

2.4.3 Reputational

- Risk – damage to corporate reputation from media and non-governmental organization campaigns, shareholder resolutions and changing customer preferences.
- Opportunity – benefits from implementing and communicating sustainable purchasing, operating, or investment practices in order to differentiate corporate brands.

In 1995, aluminium manufacturer Alcan sought to divert a river to generate hydropower for one of its smelters in Canada. However, local indigenous communities objected since the river was a source of fresh water, fish and cultural services for them. In the end Alcan was unable to receive consent and ultimately abandoned the project, losing US$500 million in upfront investment (Esty and Winston 2006).

GDF Suez aims to improve relations with stakeholders by partnering with France's National Museum of Natural History to create hundreds of kilometres of grassland corridors above and around their natural gas pipelines. These corridors in the Ile-de-France region increase the aesthetic value of lands through which the pipelines pass, and decrease operational disruption to surrounding ecosystems (WBCSD et al. 2002).

According to Steve Hounsell, Environmental Advisor of Ontario Power Generation, an electric utility, biodiversity programmes are typically very low cost relative to the benefits gained in corporate image:

> Groups that would be normally critical of OPG (and our fossil emissions) are very much 'on-side' and supportive of this programme. They have become our allies. This helps to earn a 'community license to operate' and although it is difficult to monetise, the loss of community support can, conversely, spell the demise of operations. (WBCSD 2008)

2.4.4 Market and product

- Risk – customers switching to other suppliers that offer products with lower ecosystem impacts or governments implementing new sustainable procurement policies.
- Opportunity – launching new products and services that reduce customer impacts on ecosystems, participating in emerging markets for carbon sequestration and watershed protection, capturing new revenue streams from company-owned natural assets, and offering eco-labelled wood, seafood, produce and other products.

Walmart, the world's largest retailer, is working with the Global Aquaculture Alliance (GAA) and the Aquaculture Certification Council (ACC) to certify, by 2011, that all its foreign shrimp suppliers adhere to US standards of Best Aquaculture Practices (Walmart 2010). In addition, Walmart has set itself the goal that by 2011 the company will purchase only wild-caught fresh and frozen fish for the USA from Marine Stewardship Council (MSC)-certified fisheries.

2.4.5 Financial

- Risk – higher costs of capital or difficulties acquiring debt or equity as banks and investors adopt more rigorous lending and investment policies.
- Opportunity – more favourable financing terms or improved access to capital for companies supplying products and services that improve resource efficiency or restore degraded ecosystems.

Rabobank has developed a policy that will enable the bank to exclude certain undesired operational practices in the palm oil supply chain. The policy includes binding conditions on the bank and its clients that should help promote more sustainable operations (Rabobank 2010).

Table 2.3 Biodiversity risks in selected sectors

Sector	F&C	UNEP FI (impact)	Oekom & Eurosif	Impacts	Dependence
Paper and pulp	High	Med	High	Potentially high – use of water, conversion of land, reduction of soil fertility, illegal logging	High – water, soil, pest control, yield levels, loss of raw materials
Soya	High	High	High	Potentially high – use of water, conversion of land, reduction of soil fertility	High – water, soil, pest control, yield levels
Cattle	High	High	High	Potentially high – conversion of land for feed and grazing, water usage	High – access to appropriate feed
Oil and gas	High	High	Med	Potentially high – increasingly reserves are located in sensitive sites and strong management practices are required to secure licence to operate	Moderate – level depends on location (e.g. for a refinery on coastal marsh, if the marsh is destroyed then there may be limited protection from storm surge and water)
Mining and metals	High	High	Med	Potentially high (see oil and gas)	Moderate – water (see oil and gas)
Sugarcane	High	High	High	Potentially high – conversion of land for feed and grazing, water usage	High – access to appropriate feed
Food	High	High	High	Potentially high – conversion of land for feed and grazing, water usage, over-exploitation of fish, etc.	High – water, soil, pollinators, pest control, raw materials (e.g. fish)
Beverage	Med	High	High	Potentially high – conversion of land for feed and grazing, water usage	High – water, some raw materials (e.g. agave)
Financial services	Med	N/a	N/a	Indirect impacts – although exposed to reputational risk if implicated in financing activities that destroy biodiversity (e.g. investors of APP)	Low – currently financial system does not penalize mismanagement of BES

Table 2.3 *(Cont'd)*

Sector	F&C	UNEP FI (impact)	Oekom & Eurosif	Impacts	Dependence
Pharmaceuticals and biotech	Med	N/a	N/a	Low – relatively low impact on biodiversity through extraction of natural resources; potential positive impact through equitable benefit sharing	Moderate – sourcing of natural ingredients is now rare, although concerns remain about bio-prospecting and impacts and dependence on biodiversity and ecosystem services
General retailers	Med	N/a	N/a	Indirect through supply chains	Moderate – often retailers can switch products and therefore risk exposure in terms of product cost and security of supply is relatively low; exposure to consumer pressure is increasing and may have impacts on reputation
Leisure and hotels	High	Med	High	Moderate – impact through construction, use of raw materials and sourcing of food stuff, potential cumulative impact of tourism may be damaging, release of emission from air travel also impacts biodiversity	Moderate – some elements of tourism are dependent on continued access to clean water, pristine coral reefs, pleasing surroundings
Construction and building materials	High	Med	High	High – through climate change (cement), sourcing of raw materials (e.g. timber and development of urban areas/roads)	Moderate – some dependence through timber and adaptation of networks, potential to realize opportunities through use of natural infrastructure (e.g. to ensure water quality)

Table 2.3 *(Cont'd)*

Sector	F&C	UNEP FI (impact)	Oekom & Eurosif	Impacts	Dependence
Tobacco	Med	N/a	High	Moderate – social issues overshadow environmental in terms of risk	Moderate – dependent on healthy soils for good yields and on timber for curing
Utilities and electricity	High	N/a	N/a	High – provision of water may disrupt water supplies to biodiversity, generation of clean water releases greenhouse gas emissions, conversion of habitat for hydropower	High – in particular water provision

Goldman Sachs established and funded a Center for Environmental Markets to undertake independent research with partners in the academic community and NGOs to explore and develop public policy options for establishing effective markets around climate change, biodiversity conservation and ecosystem services (Goldman Sachs 2008).

Three recent evaluations attempt to assess the level of biodiversity risk in different business sectors: (1) 'Is biodiversity a material risk?' by F&C Investments (2004); (2) 'Bloom or bust?' by UNEP FI (2008); and (3) 'Biodiversity' by Oekom Research and Eurosif (2009). The results of these assessments are summarized in Table 2.3.

2.5 Conclusion

2.5.1 There is strong evidence of global BES decline and pressures are increasing

Most regulating ecosystem services are under pressure and many ecosystems have experienced severe degradation over the past 50 years. Projections of the status of biodiversity and ecosystems by 2050 suggest that the world is likely to continue to lose biodiversity and many ecosystem services if we maintain our current development path, consumption patterns and levels of resource use.

Because all businesses depend on biodiversity and ecosystem services, either directly or indirectly, their continued decline creates real business risks. As a consequence, business should systematically review their BES dependence, covering direct operations and those of suppliers and customers. Business should also assess their impacts on nature, direct and indirect, positive or negative.

2.5.2 Business should make use of integrated environmental management and assessment tools

It is important that businesses better understand their dependence and impacts on biodiversity and ecosystems, and that they design appropriate mitigation and management responses. Biodiversity and ecosystem values can and should be integrated more fully into corporate decision making at all levels. BES impacts and dependence can affect the competitive position and performance of a company. In addition, public perception of business impacts on nature may influence consumer trends, corporate reputation and/or the ability to maintain the legal or social licence to operate.

2.5.3 Many of the drivers of biodiversity loss and ecosystem decline pose challenges for business

Fresh water scarcity, habitat change and degradation, climate change, pollution, over-exploitation and the spread of invasive alien species can all compromise business operations and investments. Measures to reduce the direct drivers of BES decline should continue and may be supported by business. This includes developing appropriate corporate strategy, policy and operational responses, guided by the hierarchy of avoiding, minimizing, mitigating and (where feasible) offsetting negative impacts.

It is also clear that BES loss cannot be tackled by business alone, but requires partnerships involving governments, business and civil society. Impacts on BES need to be understood on both a local and a global scale. While the legal frameworks in which companies do business are primarily local, the world is increasingly global in its economics and in the spread of information. Meeting growing demand for commodities like food and water while also maintaining and restoring BES is likely to involve changes in public policies and regulations, as well as business practices. There is a need for better technology and business can play a key role as a solutions provider.

2.5.4 Halting biodiversity loss and ecosystem degradation presents new business opportunities

Business can prosper by developing technologies and products that will serve as substitutes, reduce degradation, restore ecosystems or increase the efficiency of natural resource use. Other opportunities include new markets such as water quality trading, certified sustainable products, wetland and species banking or other new revenue streams for assets that are currently unrealized.

Acknowledgements

Adachi Naoki (Response Ability), Alistair McVittie (SAC), Delia Shannon (Aggregate Industries), Gerard Bos (Holcim), Luke Brander (IVM), Richard Mattison (Trucost plc), Alison Reinert (Syngenta), Donn Waage (NFWF), Gigi Arino (Syngenta), Jeffrey Wielgus (WRI), Jennifer Shaw (Syngenta), JiSu Bang (Syngenta), Juan Valero-Gonzalez (Syngenta), Juan Carlos Vasquez (CITES Secretariat), Rufus Isaacs (Michigan State University), Steve Bartell (E2 Consulting).

References

AgMRC (2010) 'Blueberries' Agricultural Marketing Resource Center. URL: www.agmrc.org/commodities__products/fruits/blueberries.cfm

Alfa Laval (2010) URL: www.pureballast.alfalaval.com/PureBallast.aspx (last accessed 28 May 2010).

BASF (n.d.) 'Rawcliffe Bridge: The partnership – commercial farming and biodiversity – five successful seasons'. URL: www.agro.basf.com/agr/AP-Internet/en/function/conversions:/publish/upload/sustainability/Biodiversity_Rawcliffe_Inners.pdf (last accessed 2 February 2011).

Batker, D.P., de la Torre, I., Costanza, R., Swedeen, P., Day, J.W., Jr., Boumans, R. and Bagstad, K. (2010) *Gaining Ground – Wetlands, Hurricanes and the Economy: The Value of Restoring the Mississippi River Delta*. Earth Economics, Tacoma, WA.

Boyd, J. (2010) *Lost Ecosystem Goods and Services as a Measure of Marine Oil Pollution Damages*. DP 10–31. Resources for the Future, Washington, DC.

BP (2010) 'Group results: Second quarter and half year 2010'. URL: www.bp.com/liveassets/bp_internet/globalbp/STAGING/global_assets/downloads/B/bp_second_quarter_2010_results.pdf (last accessed 20 October 2010).

Brander, L. (2010) Personal communication, Institute for Environmental Studies, VU University Amsterdam.

Butchart, S. H., Walpole, M., Collen, B., van Strien, A., Scharlemann, J. P., Almond, R.E., et al. (2010) 'Global biodiversity: indicators of recent declines', *Science* 328 (5982), 1164–8.

CBD (2010) *Global Biodiversity Outlook 3*, Montréal. Secretariat of the Convention on Biological Diversity. URL: www.cbd.int/doc/publications/gbo/gbo3-final-en.pdf

Chevassus-au-Louis, B., Salles, J.-M. and Pujol, J.-l. (2009) *Approche économique de la biodiversité et des services liés aux écosystèmes: Contribution à la décision publique*. Centre d'Analyse Stratégique. Report to the Prime Minister (Government of France). URL: www.strategie.gouv.fr/IMG/pdf/rapport_bio_v2.pdf (last accessed 6 November 2009).

Cleveland, C. (2010) *Deepwater Horizon Oil Spill*. The Encyclopedia of Earth. URL: www.eoearth.org/article/Deepwater_Horizon_oil_spill (last accessed 25 October 2010).

Colipa (2008) *Value & Values: In Today's Cosmetics Industry, Annual Report 2008*, European Cosmetics Association, Brussels. URL: www.colipa.eu/downloads/16.html

Costanza, R., Batker, D., Day, J., Feagin, R.A., Martinez, M.L. and Roman, J. (2010) *The Perfect Spill: Solutions for Averting the Next Deepwater Horizon*. URL: www.thesolutionsjournal.com/node/629 (last accessed 25 October 2010).

Coulson, A.B. (2009) 'How should banks govern the environment? Challenging the construction of action versus veto'. *Business Strategy and the Environment* (March) 18 (3), 149–61.

Environmental Justice Foundation (2005) *White Gold: The True Cost of Cotton*. Environmental Justice Foundation, London, UK.

Esty, D. and Winston, A. (2006) *Green to Gold: How Smart Companies Use Environmental Strategy to Innovate, Create Value, and Build Competitive Advantage*. New Haven, CT, Yale University Press.

F&C Investments (2004) 'Is biodiversity a material risk for companies? An assessment of the exposure of FTSE sectors to biodiversity risk', F&C Management Limited, London. URL: www.businessandbiodiversity.org/pdf/FC%20Biodiversity%20Report%20FINAL.pdf.

FAO (1998) *Time to Save the Aral Sea*? Agriculture and Consumer Protection Department, Paris.

FAO (2001) *Global Forest Resources Assessment 2000*. FAO, Rome. URL: ftp://ftp.fao.org/docrep/fao/003/y1997E/frA%202000%20Main%20report.pdf

FAO (2007) *The World's Mangroves 1980–2005*. FAO Forestry Paper, Rome. URL: ftp://ftp.fao.org/docrep/fao/010/a1427e/a1427e00.pdf

FAO (2009) *The State of World Fisheries and Aquaculture – 2008*. FAO, Rome. URL: ftp://ftp.fao.org/docrep/fao/011/i0250e/i0250e.pdf

FSC (2005) *Leading Our World Towards Responsible Forest Stewardship: A Progress Report*: Forest Stewardship Council, Bonn. URL: www.fsc.org/fileadmin/webdata/public/document_center/publications/annual_reports/FSC_GA2005_Brochure_Lo wRes.pdf

Gallai, N., Salles, J.M., Settele, J. and Vaissière, B.E. (2009) 'Economic valuation of the vulnerability of world agriculture confronted with pollinator decline'. *Ecological Economics* 68 (3), 810–21.

Glazovsky, N.F. (1990) 'Ideas on an escape from the "Aral Crisis"'. *Soviet Geography* 32 (2), 73–89.

Glazovsky, N.F. (1995) *The Aral Sea Basin. Regions at Risk: Comparison of Threatened Environments*, United Nations University Press, Tokyo and New York.

Global Environmental Management Initiative (2002a) 'Connecting the drops toward creative water strategies: A water sustainability tool, Anheuser-Busch Inc. – Exploring water connections along the supply chain'. Washington, DC. URL: www.gemi.org/water/anheuser.htm

Global Environmental Management Initiative (2002b) 'Connecting the drops toward creative water strategies: A water sustainability tool, DuPont: Managing strategic risk through innovative wastewater treatment'. Washington, DC. URL: www.gemi.org/water/dupont.htm

Goldman Sachs (2008) *Environmental Report*. URL: www2.goldmansachs.com/services/advising/environmental-markets/documents-links/env-report-2008.pdf

Goldman Sachs, Environmental Policy Framework. URL: www2.goldmansachs.com/services/advising/environmental-markets/documents-links/environmental-policyframework.pdf

Grangeon, D. and Cousin, P. (2009) 'Toward a greener road pricing system in Europe'. European Transport Conference, Netherlands (October).

Hanson, Craig, Janet Ranganathan, Charles Iceland, and John Finisdore (2008) *The Corporate Ecosystem Services Review: Guidelines for Identifying Business Risks and Opportunities Arising from Ecosystem Change*. WRI, WBCSD and Meridian Institute, Washington, DC. URL: http://pdf.wri.org/corporate_ecosystem_services_review.pdf

Houdet, J. (2010) *The Financial Impacts of BP's Response to the Deepwater Horizon Oil Spill. Summary in English for TEEB*. (Original available in French only) URL: www.synergiz.fr/wp-content/uploads/2010/12/Etude-cas_BP_implications_dommages_dec2010.pdf

HSBC (2005) 'Freshwater infrastructure sector guideline'. URL: www.hsbc.com/1/PA_1_1_S5/content/assets/csr/freshwater_infrastructure_guideline.pdf (last accessed 15 June 2010).

IFAW (2009) *Whale Watching Report*. International Fund for Animal Welfare.

IOPC Funds (2009) *Annual Report 2008*, International Oil Pollution Compensation Funds, London.

IPCC (2007) *Fourth Assessment Report Climate Change, Synthesis Report*. Geneva. URL: www.ipcc.ch/publications_and_data/publications_ipcc_fourth_assessment_report_synthesis_report.htm

Kijne, J.W. (2005) *Aral Sea Basin Initiative: Towards a Strategy for Sustainable Irrigated Agriculture with Feasible Investment in Drainage*. Synthesis report. FAO, Rome.

L'Oréal (2009) *2008 Sustainable Development Report*. URL: www.loreal.com/_en/_ww/pdf/LOREAL_RDD_2008.pdf

Lang, G. (2002) 'Deforestation, floods, and state reactions in China and Thailand', Working Paper Series, No. 21, City University of Hong Kong, SEARC.

Larsen, J. (2002) *Illegal Logging Threatens Ecological and Economic Stability*, Earth Policy Institute. URL: www.earth-policy.org/index.php?/plan_b_updates/2002/update11 (last accessed 28 May 2010).

Lovell, S.J. and Stone, S.F. (2005) 'The economic impacts of aquatic invasive species: A review of the literature', National Center for Environmental Economics, Working Paper Series, No. 05-02, U.S. EPA, Washington, DC.

MA (2005a) *Ecosystems and Human Well-being: Biodiversity Synthesis*. World Resources Institute. Island Press, Washington, DC. URL: www.millenniumassessment.org/documents/document.354.aspx.pdf (last accessed 23 June 2010).

MA (2005b) *Ecosystems and Human Well-being: Opportunities and Challenges for Business and Industry*. Island Press, Washington, DC. URL: www.millenniumassessment.org/documents/document.353.aspx.pdf

MA (2005c) *Ecosystems and Human Well-being: Current State and Trends – Findings of the Condition and Trends Working Group*. Hassan, R., Scholes, R. and Ash, N. (eds) URL: www.millenniumassessment.org/en/Condition.aspx

MA (2005d) 'Ecosystems and Human Well-being: Scenarios – Findings of the Scenarios Working Group'. URL: www.millenniumassessment.org/en/Scenarios.aspx

Mardas, N., Mitchell, A., Crosbie, L., Ripley, S., Howard, R., Elia, C. and Trivedi, M. (2009) 'Global forest footprints', Forest footprint disclosure project, Global Canopy Programme, Oxford. URL: http://forestdisclosure.com/docs/FFD-Global-Forest- Footprints-Report.pdf

McVittie, A. (2010) Personal communication, Land Economy & Environment Research Group, Scottish Agricultural College.

Micklin, P.P. (1992) 'The Aral crisis: Introduction to the special issue'. *Post-Soviet Geography* 33 (5), 269–82.

Micklin, P. and Aladin, N.V. (2008) 'Reclaiming the Aral Sea'. *Scientific American*, April. URL: www.scientificamerican.com/article.cfm?id=reclaiming-the-aral-sea (last accessed 28 May 2010).

Mulder, I. and Koellner, T. (2011) 'Hardwiring green – how banks account for biodiversity risks and opportunities'. *Journal of Sustainable Finance and Investment* 1 (2).
Natura (2008) *Annual Report 2008*. URL: www2.natura.net/Web/Br/relatorios_anuais/_PDF/AnnualReport2008.pdf
Newman, D.J. and Cragg, G.M. (2007) 'Natural products as sources of new drugs over the last 25 years'. *Journal of Natural Products* 70, 461–77.
OECD (2008) *Environmental Outlook to 2030*. Organisation for Economic Co-operation and Developement, Paris. URL: www.oecd.org/document/20/0,3343,en_2649_34305_39676628_1_1_1_37465,00.html
Oekom Research and Eurosif (2009) *Biodiversity: Theme report – second in a series*. URL: www.oekom-research.com/homepage/eurosif-sr_biodiversity.pdf (last accessed 2 February 2011).
Olsen, N. and Shannon, D. (2010) 'Valuing the net benefits of ecosystem restoration: the Ripon City Quarry in Yorkshire', Ecosystem Valuation Initiative, Case Study No. 1, WBCSD and IUCN, Geneva and Gland.
Peters, J., Shaw, J., Valero-Gonzalez, J., Arino, G., Bang, J., Reinert, A., et al. (2010) 'Operation Pollinator: Investing in natural capital for agriculture' (case study for TEEB).
Pettis, J.S. and Delaplane, K.S. (2010) 'Coordinated responses to honey bee decline in the USA'. *Apidologie* 41 (3), 256–63.
PRI UNEP FI (2011) *Universal Ownership: Why Environmental Externalities Matter to Institutional Investors*, PRI Association and UNEP Finance Initiative.
Rabobank (2010) 'Rabobank's position on palm oil'. URL: www.rabobank.com/content/images/positionpaper_palmoil_tcm43-107432.pdf (last accessed 15 June 2010).
Ricoh (2009) *Group Sustainability Report (Environment)*, p. 70. URL: www.ricoh.com/environment/report/pdf2009/all.pdf (last accessed 4 July 2010).
Riedmiller, S. (2010) Pers. comm.
Sampath, P.G. (2005) *Regulating Bioprospecting: Institutions for Drug Research, Access and Benefit-Sharing*. United Nations University Press, Tokyo and New York.
Seneca Creek Associates (2004) '"Illegal" logging and global wood markets: The competitive impacts on the U.S. wood products industry'. URL: www.illegal-logging.info/uploads/afandpa.pdf
Severskiy, I., Chervanyov, I., Ponamorenko, Y., Novikova, N.M., Miagkov, S.V., et al. (2005) *Global International Waters Assessment (GIWA) 24*, Aral Sea. University of Kalmar, Sweden, UNEP.
Song, C. and Zhang, Y. (2010) Forest cover in China from 1949 to 2006, in Nagendra, H. and Southworth, J. (eds) *Reforesting Landscapes: Linking Pattern and Process*, Landscape Series 10. Springer, Dordrecht, Heidelberg, London, New York.
Spurgeon, J.P.G. (2006) 'Reefs in an economics context: Time for a "Third Generation" economics based approach to coral management', in Cote, I.M. and Reynolds, J.D. (eds) *Coral Reef Conservation*. Cambridge University Press, Cambridge.
Stenek, V., Colley, M., Connell, R. and Firth, J. (2011) *Climate Risk and Business: Agribusiness – Ghana Oil Palm Development Company*, International Finance Corporation, World Bank Group.
Sun, J., Zhao, C. and Wang, L. (2002) *The Long March of Green: The Chronicle of Returning Agricultural Land to Forests in China*. China Modern Economics Press, Beijing, P.R. China.
Sutherland, P.D. (2007) *Major Water Resource Challenges – A View Across Two Southeast States*, Proceedings of Ozwater Conference, March, Sydney, Australia.
TEEB Foundations (2010) *The Economics of Ecosystems and Biodiversity: Ecological and Economic Foundations*. (ed. P. Kumar), Earthscan, London.
TEEB in Policy Making (ed.) (2011) *The Economics of Ecosystems and Biodiversity in National and International Policy Making* (ed. P. ten Brink), Earthscan, London.
Temirov, R. (2003) 'Lobbying grows in Moscow for Siberia–Uzbekistan water scheme', Eurasianet.org.
UNEP – (2008) 'Environment Alert Bulletin: Coastal degradation leaves the Caribbean in troubled waters'. URL: www.grid.unep.ch/product/publication/download/ew_caribbean_runoffs.en.pdf (last accessed 18 May 2008).
UNEP (2010) UNEP *Emerging Issues: Global Honey Bee Colony Disorder and Other Threats to Insect Pollinators*, United Nations Environment Programme, Nairobi.
UNEP FI (2008) *Biodiversity and Ecosystem Services: Bloom or Bust? A Document of the UNEP FI Biodiversity & Ecosystem Services Work Stream*. URL: www.unepfi.org/fileadmin/documents/bloom_or_bust_report.pdf (last accessed 2 February 2011).
UNEP-WCMC – United Nations Environment Programme World Conservation Monitoring Centre (2002) *World Atlas of Biodiversity: Earth's Living Resources for the 21st Century*. URL: www.unep-wcmc.org/information_services/publications/biodiversityatlas/presspack/press/release.htm

Union for Ethical BioTrade (2010) Biodiversity Barometer 2010. URL: www.countdown2010.net/2010/wp-content/uploads/UEBT_BIODIVERSITY_BAROMETER_web-1.pdf

United States Department of Agriculture (2008) *Cotton: Production, Supply and Distribution*, Foreign Agricultural Service.

UNWWAP (2003) *Water for People, Water for life*. United Nations World Water Assessment Programme URL: www.unesco.org/water/wwap/wwdr/wwdr1/

USDA NASS (2008) Noncitrus Fruits and Nuts. US Department of Agriculture, National Agricultural Statistics Service. URL: www.nass.usda.gov/

vanEngelsdorp, D., Hayes, J.Jr., Underwood, R.M., Caron, D. and Pettis, J. (2011) 'A survey of managed honey bee colony losses in the USA, fall 2009 to winter 2010'. *Journal of Apicultural Research* 50 (1), 1–10. URL: www.ibra.org.uk/articles/US-bee-loss-survey-2009-10

Walmart (2010) *Global Sustainability Report: 2010 Progress Update, We Save People Money so They Can Live Better*. URL: http://cdn.walmartstores.com/sites/sustainabilityreport/2010/WMT2010GlobalSustainabilityReport.pdf

Wang Hongchang (1997) *Deforestation and Desiccation in China: A Preliminary Study, in Mao Yu-shi, Ning Datong, Xia Guang, Wang Hongchang and Vaclav Smil*, An assessment of the Economic Losses Resulting from Various Forms of Environmental Degradation in China, Occasional Paper of the Project on Environmental Scarcities, State Capacity, and Civil Violence. American Academy of Arts and Sciences, Cambridge, and University of Toronto, URL: www.library.utoronto.ca/pcs/state/chinaeco/forest.htm (last accessed 6 July 2010). Also published in Smil, Vaclav and Mao Yushi (1998) The Economic Costs of China's Environmental Degradation. American Academy of Arts and Sciences, Cambridge. URL: www.amacad.org/publications/china2.aspx (last accessed 6 July 2010).

WBCSD (2007) *The Sustainable Forest Products Industry, Carbon and Climate Change: Key Messages for Policy-Makers*, World Business Council for Sustainable Development, Geneva. URL: www.wbcsd.org/DocRoot/oNvUNPZMuugn75jrL8KS/sfpicarbon-climate.pdf

WBCSD (2008) 'Gaz de France: Partnering for conservation', World Business Council for Sustainable Development, Geneva. URL: www.wbcsd.org/DocRoot/I1nBxnCY gi4PKwa5j8Tk/GazdeFrancefullcasefinal.pdf

WBCSD, IUCN and Earthwatch (2002) 'Business and biodiversity: Handbook for corporate action, World Business Council for Sustainable Development', Geneva. URL: www.wbcsd.org/web/publications/business_biodiversity2002.pdf

WBCSD, WRI, IUCN and Earthwatch (2006) 'Business and Ecosystems, Issue Brief: Ecosystem challenges and business implications', Geneva. URL: www.wbcsd.org/DocRoot/Ejk5KCJOlkVkRngCksWD/Business%20and%20Ecosystems_211106_final.pdf

Werf, G.R. van der, Morton, D.C., DeFries, R.S., Olivier, J.G.J., Kasibhatla, P.S., Jackson, R.B., Collatz, G.J. and Randerson, J.T. (2009) 'CO_2 emissions from forest loss'. *Nature Geoscience* 2, 737–8.

Wilkinson, C. (2008) *Status of Coral Reefs of the World: 2008*. Global Coral Reef Monitoring Network and Reef and Rainforest Research Centre, Townsville, Australia.

WTTC (2010) *Travel and Tourism Economic Impact 2010*. World Travel and Tourism Council.

Xu, J.T. and Cao, Y.Y. (2002) 'Converting steep cropland to forest and grassland: Efficiency and prospects of sustainability'. *International Economic Review* (Chinese), 2, 56–60.

Yin, R.S. (1998) 'Forestry and the environment in China: The current situation and strategic choice'. *World Development* 26 (12), 2153–67.

Young, L. (2003) 'Putting an economic value on environmental or natural benefits that create commercial wealth is a concept that is gaining momentum'. *The Source*, A Magazine by Melbourne Water, Issue 26 (June).

Annex 2.1 Case studies: Cotton, the Aral Sea and timber in China

Authors: Mark Trevitt (Trucost plc), Alistair McVittie (Scottish Agricultural College), Luke Brander (Institute for Environmental Studies), Joshua Bishop (IUCN)

Objectives and methodology

This annex examines the economic impacts and dependence of business on ecosystems and biodiversity through case studies of the agriculture and textile industry in Central Asia and the construction and materials sector in China. The case studies show how unsustainable use of ecosystem services and failure to account for the non-market values of ecosystems can lead to environmental crises, with profound economic consequences and impacts on the business bottom line.

Both case studies examine the costs resulting from ecosystem degradation that may be embedded in a company's supply chain, due to the impact and dependence of the raw materials they use on particular ecosystem services. In both instances, the economic consequences of ecosystem degradation are highlighted alongside the value of the ecosystem services lost as a result of failing to use natural resources sustainably. Due to the lack of primary data, benefits transfer techniques are employed to estimate the value of ecosystem services lost and to illustrate how economic valuation of the benefits provided by ecosystems can help safeguard important business values for the future.

Case Study 1: Cotton production and the destruction of the Aral Sea

A striking example of how the unsustainable use of scarce water resources can destroy an entire ecosystem is provided by the desiccation of the Aral Sea. Situated between the nations of Kazakhstan and Uzbekistan in central Asia (formerly part of the Soviet Union), the Aral Sea was the world's fourth largest inland sea in 1960, providing a wealth of ecosystem services to surrounding communities. By 2007, the Aral Sea had shrunk to 10 per cent of its original size, mainly due to water abstraction from its two major tributaries, the Amu Darya River and the Syr Darya River (Micklin and Aladin 2008). The diversion of water from these rivers was in turn a direct result of the development of irrigated cotton production in the surrounding region.

The crisis had its roots in the early decades of the twentieth century, when the government of the Soviet Union initiated a plan to expand irrigation in the region, to cultivate cotton for export, and thereby to increase the standard of living of the region's growing population (Nalwalk 2000). The government recognized that the expansion of irrigation for the cultivation of cotton would reduce water inflow to the Aral Sea, but the trade-off was considered worthwhile, as it was believed that 'a cubic meter of river water used for irrigation would be more economically beneficial than the same volume delivered to the Aral Sea' (Micklin 1988).

In 1956, the Kara Kum Canal was opened, resulting in the diversion of large amounts of water from the Amu Darya River. The subsequent reduction in river water flow ultimately resulted in the separation of the Aral Sea into two water bodies in 1987 (a small Aral Sea in the north and a large Aral Sea in the south), as well as significant increases in the salinity of the Aral Sea.

The expansion of irrigated area and consumption of water resources was driven largely by growth in cotton production. In recent years the greatest consumer of fresh water in the Aral Sea basin has been Uzbekistan, using on average about 54 per cent of the region's total water resources (UNEP 2006). In 1991, cotton accounted for more than 65 per cent of Uzbekistan's gross domestic output, consumed 60 per cent of its resources and employed 40 per cent of the country's labour force, while more than 70 per cent of arable land in the republic was devoted to cotton production (Nalwalk 2000).

Over the period 1960–1990, the development of irrigation around the Aral Sea increased from approximately 4.5 million ha to just over 7 million ha, while the surface area of the Aral Sea declined from almost 70km^2 to under 40km^2 (see Figure 2A.1). While some economic progress was achieved in the short run, it was realized at the expense of the environment and long-term economic sustainability.

Impacts on ecosystems and people

Increased use and runoff of pesticides and fertilizer resulted in the pollution of surface and groundwater, while declining downstream water availability and increased salinity deprived the region's lakes and wetlands of their life source. As a result, the ecosystems of the Amu Darya delta in Uzbekistan and the Syr Darya delta in Kazakhstan have suffered substantial damages. In the Amu Darya delta, wetlands that covered some 550,000ha in 1960 were reduced by 95 per cent to about 27,500ha in 1990, replaced by sandy deserts, while more than 50 delta lakes, covering some 60,000ha, simply dried up (FAO 1998). Similarly, the

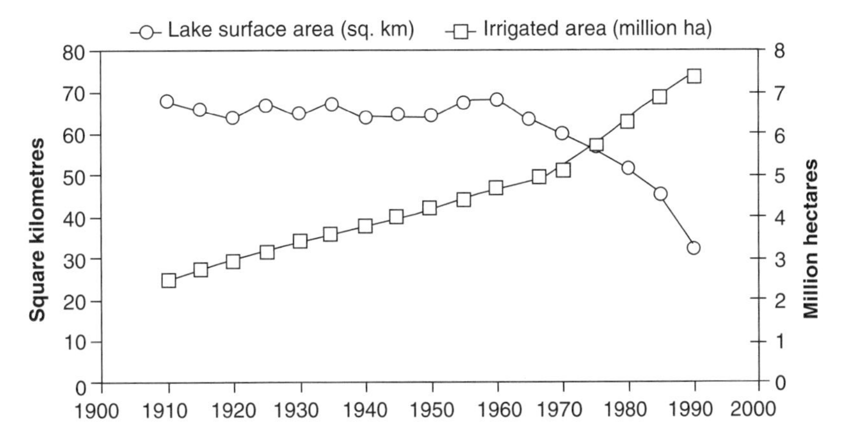

Figure 2A.1 *Irrigated area versus total surface area of the Aral Sea*

Source: Cai et al. (2003) based on Micklin (1993)

lakes of the Syr Darya delta shrank from about 500km^2 in 1960 to 40km^2 in 1980 (Micklin 1992). So-called Tugai forests, which covered about 100,000 hectares in the Amu Darya delta in 1950, were reduced to just 20,000–30,000 hectares by 1999 (Severskiy et al. 2005). Other impacts of water diversion and pollution are summarized below:

- Before 1960, over 70 species of mammals and 319 species of birds lived in the river delta; by 2007 only 32 species of mammals and 160 species of birds remained (Severskiy et al. 2005).
- The number of fish species in the lakes dropped from 32 to 6, due to increased salinity and loss of spawning and feeding grounds (Micklin and Aladin 2008).
- Commercial fisheries that produced some 40,000 metric tons of fish in 1960 were wiped out by the mid-1980s, with the loss of over 60,000 jobs (Micklin and Aladin 2008).
- Poor water management and derelict infrastructure led to declining soil fertility and soil erosion, threatening 19 per cent of irrigated land (World Bank 2003).
- With the shrinking of the Aral Sea, the climate of the surrounding region has become more continental, with shorter, hotter, drier summers and longer, colder, snowless winters. The growing season has been reduced to an average of 170 days per year, while dust storms occur on more than 90 days per year, on average (FAO 1998).
- Over 15 years there was a 3,000 per cent increase in reported chronic bronchitis and kidney and liver diseases, including cancer, while arthritic diseases increased by 6,000 per cent. The infant mortality rate is among the world's highest (FAO 1998).
- Average life expectancy in surrounding populations decreased from 65 to 61 years (Micklin and Aladin 2008).

Focus on the agriculture and textile industry

The case of the Aral Sea can be understood as a creeping environmental problem, where changes accumulate over time, degradation is generally imperceptible and the full scale of the impact is not recognized until a crisis occurs. One early indicator of environmental deterioration in the basin was a decline in cotton yields, due to water quality and soil problems linked to irrigation (see Figure 2A.2).

In response to soil degradation, Uzbek farmers increased the volume of water applied, 'flushing' their fields with irrigation water to wash away excess salt. This practice has threatened the survival of Uzbek agricultural production, especially cotton, as rising salt levels kill or retard the growth of natural vegetation and crops (Spoor and Krutov 2004).

No previous comprehensive estimates were found of the economic damages resulting from the desiccation of the Aral Sea. One study, published in 1990, examined the cost of measures to redress some negative environmental consequences, suggesting that the minimum damage to the environment as a result of unsustainable agricultural and irrigation practices in the Aral Sea was at least US$1.4 billion, based on estimates of the cost of measures to prevent polluted drainage water from entering rivers, reconstruction of the irrigation system, introduction of new plants and irrigation techniques, and stabilization of the sea floor (Glazovsky 1990, 2005). In addition, the

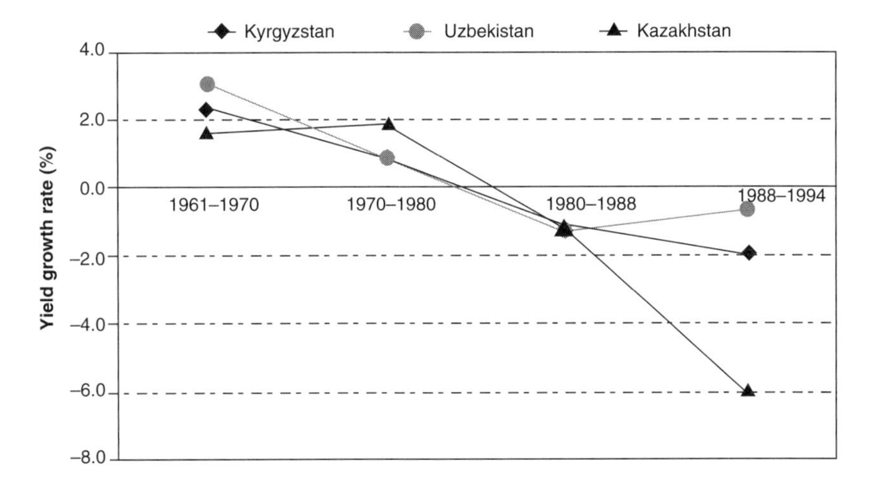

Figure 2A.2 *Cotton yield trend in three Central Asian countries near the Aral Sea*

Source: Cai et al. (2003) based on FAOSTAT (www.fao.org)

cost to improve sanitary, hygiene and medical services, create new jobs and reform the economy of the area was estimated at over US$3.49 billion (Glazovsky 1995). Other studies examine the cost of constructing two canals to redirect water from the Volga, Ob and Irtysh rivers in order to restore the Aral Sea to its former size over a 20–30-year period, estimated at over US$30 billion (Temirov 2003).

None of the studies cited above explicitly consider the loss of ecosystem services in the Aral Sea basin. For this case study, a meta-analytic value function for wetlands was applied, based on previously estimated parameters for spatial variables, size, type, abundance, GDP per capita and population density (Brander et al. 2006; Ghermandi et al. 2009). This function was used to value the loss of ecosystem services resulting from the disappearance of 522,500ha of wetlands over the period 1960–1990, suggesting annual economic losses of around US$100 million (Brander 2010). A summary of the estimation procedure and results are provided in Table 2A.1.

This analysis covers only a portion of the overall ecosystem service losses and may underestimate the true scale of losses, given that the transfer of monetary values from other wetland ecosystems may not reflect the specific conditions of the Aral Sea basin. Nevertheless, from this analysis it seems clear that the externalities associated with increased cotton production in the Aral Sea basin are significant. While the agricultural industry has not borne all of these costs, it has been severely affected. Unsustainable irrigation practices, resulting in water-logging, increased soil salinity and lower yields, have reduced the value of crop production by US$1.4 billion per year, or approximately one-third of the value of potential output (Glazovsky 1990).

Implications for business

The manufacture of cotton garments by the retail clothing industry is connected to a chain of environmental dependence and impacts on water resources in the countries where cotton is grown and processed – mainly involving water abstraction and pollution during cultivation and processing. Almost two-thirds of all cotton produced worldwide is used by the textiles industry for clothing manufacture (Chapagain et al.

Table 2A.1 Estimation of aral sea wetland values

Explanatory variables	Variable label	Estimated coefficient	P value	Signif.	Aral Sea wetland parameter values	Product of coefficient and Aral Sea wetland parameter values	Notes & assumptions
	(Constant)	−0.970	0.709		1.00	−0.97	Adjusted $R^2 = 0.37$
Valuation method							
Contingent valuation	CVM	0.317	0.625		1.00	0.32	Yes
Hedonic pricing	HP	−2.328	0.043	**	0.00	0.00	No
Travel cost	TCM	−0.705	0.261		0.00	0.00	No
Replacement cost	REPLCOST	−0.383	0.538		0.00	0.00	No
Net factor income	NFINCOME	−0.125	0.843		0.00	0.00	No
Production function	PRODFUNC	−0.091	0.896		0.00	0.00	No
Market price	MKTPRICE	−0.215	0.712		0.00	0.00	No
Opportunity cost	OPPCOST	−1.164	0.165		0.00	0.00	No
Choice model	CHOICE	−0.524	0.581		0.00	0.00	No
Marginal value	MARGINAL	0.828	0.053	**	0.00	0.00	No
Wetland type							
Inland marsh	EEA_INLND	−0.211	0.726		1.00	−0.21	Yes
Peatbog	EEA_PTBGS	−2.266	0.004	***	0.00	0.00	No
Salt marsh	EEA_SLTMR	0.073	0.901		0.00	0.00	No
Intertidal mudflat	EEA_INTRT	−0.239	0.672		0.00	0.00	No
Wetland size (ha)	LN_SIZE	−0.218	0.000	***	8.91	−1.94	See note 1
Ecosystem service							
Flood control	FLOOD	0.626	0.169		0.15	0.09	See note 2
Water supply	WATSUPP	−0.106	0.828		1.00	−0.11	Yes
Water quality	WATQUAL	0.514	0.288		1.00	0.51	Yes
Habitat and nursery	HABITAT	0.042	0.917		1.00	0.04	Yes
Recreational hunting	HUNTING	−1.355	0.002	***	1.00	−1.35	Yes
Recreational fishing	FISHING	−0.119	0.786		0.00	0.00	No
Materials	MATERIAL	−0.153	0.732		1.00	−0.15	Yes
Fuel wood	FUELWOOD	−0.959	0.198		1.00	−0.96	Yes
Non-consumptive recreation	RECREATION	0.218	0.626		0.00	0.00	No
Amenity	AMENITY	0.432	0.370		0.00	0.00	No
Biodiversity	BIODIVER	1.211	0.012	**	1.00	1.21	Yes
Socio-economic characteristics							
GDP per capita	LN_GDPPC	0.430	0.004	***	8.86	3.81	See note 3
Population within 50km	LN_POP50	0.503	0.000	***	12.24	6.16	See note 4

Table 2A.1 Estimation of aral sea wetland values *(Cont'd)*

Explanatory variables	Variable label	Estimated coefficient	P value	Signif.	Aral Sea wetland parameter values	Product of coefficient and Aral Sea wetland parameter values	Notes & assumptions
Wetland abundance							
Area of wetland within 50km	LN_WETL50	–0.125	0.118		11.08	–1.39	See note 5
					Estimated value (natural log)	5.06	
					Base value (USD/ha/yr)	157	1960 wetland abundance
					1990 value (USD/ha/yr)	229	1990 wetland abundance
					Avg value (USD/ha/yr)	**193**	
					1960 area (ha)	**550,000**	
					1990 area (ha)	**27,500**	5 per cent of 1960
					Loss in area (ha)	**–522,500**	From 1960 to 1990
					Value of loss (USD/yr)	–100,847,624	See note 6

Notes:

1) Based on the mean size of all wetlands in the underlying meta-analytic value function (sample size 222), due to lack of site-specific data.

2) Based on the mean value of flood control for all wetlands in the underlying function, due to lack of site-specific data.

3) Mean of Kazakhstan (2009 GDP/capita US$11,434 at PPP) and Uzbekistan (2009 US$2,634 GDP/capita at PPP).

4) Mean population density (pop. in 2000 (41,800,000) / total basin area (1,585,000km^2) = 26.37), multiplied by area within 50km radius (7,854km^2).

5) Wetland abundance in 1960 (mean sample value 64,860ha).

6) Annual flow of value that would have been provided by the total wetland area that was lost between 1960 and 1990.

2006). Global demand for cotton has increased steadily and, in 2008, annual world cotton production reached over 26 million tonnes (USDA 2008).

Cotton accounts for about 2.6 per cent of global freshwater consumption, or over 250 billion m^3 of water per year (Chapagain et al. 2006). Cotton is a water-intensive crop, requiring about 11,000 litres of water per kilogramme of final cotton textile, on average, worldwide (Chapagain et al. 2006). In Uzbekistan, cotton production is generally even more water-intensive, with almost 20,000 litres of water used for every kilogram of cotton harvested, due to inefficient irrigation practices, implying total consumption for cotton production at over 8.5 billion m^3 of water per year (Chapagain et al. 2006; EJF 2005).

Due to limited domestic capacity for textile production in Uzbekistan, over 70 per cent of Uzbek cotton or around 800,000 tonnes is sold on the world market every year, making it the world's second largest exporter (EJF 2005). According to the UN, the single largest consumer of Uzbek cotton is the European Union, which absorbs 29 per cent of Uzbek cotton exports, valued at around US$350 million per year (EJF 2005).

Case Study 2: Deforestation and the construction industry in China

Forests provide a range of goods and services on which societies depend (Salim and Ullsten 1999). As the world's largest country by population and third-largest country by area, China's use of forest resources affects not only the country itself but the global environment as well. When the People's Republic of China was founded in 1949, the country was very poor, damaged by years of warfare and needed major economic reconstruction. Demand for timber for construction and other uses was and has remained high.

At mid-century, China had large tracts of natural forests in the northeast (including Heilongjiang, Jilin and eastern Inner Mongolia), the southwest (covering Yunnan, western Sichuan and eastern Tibet) and parts of Xinjiang in the northwest and Hainan in the south (CNFCM 2000). During the 1950s and 1960s, however, nearly a million workers moved into forested areas to produce timber to meet the growing demand for construction materials (Xhao and Shao 2002). From the 1950s, timber harvests in China increased from about 20 million m^3 per year to some 67.7 million m^3 in 1995 (Zhang et al. 2000; Cohen and Vertinsky 2002). Demand for timber was largely driven by the construction boom; average wood consumption between 1983 and 1997 was as follows: construction (64 per cent), furniture (13 per cent), fuel (8 per cent), pulp (7 per cent) and other uses (8 per cent) (CNFCM 2000).

Impacts on forest ecosystems

Over the period 1949 to 1981, China's use of forest resources almost completely depleted its natural forest stocks. The cumulative area harvested was 75 million ha, of which 92 per cent were natural forests (Song and Zhang 2010). Rapid deforestation led to adverse structural changes, such as: (1) reduced stocking volume (m^3/ha), (2) timber age structure tilted towards younger stands, (3) changes in species composition, (4) reduced natural regeneration, and (5) low growth and yields of forest plantations (Yin 1998). The ecological functions of forest ecosystems, particularly watershed protection and soil conservation, were compromised. The decline of forest area also contributed to biodiversity loss, due to the disturbance, conversion and fragmentation of habitats (Studley 1999).

The prolonged deforestation and degradation of natural forest ecosystems in China reached a tipping point in the late 1990s, precipitating a series of ecological disasters. In 1997, severe droughts caused the lower reaches of the Yellow River to dry up for 267 days, putting industrial, agricultural and residential water uses throughout the northern plains in jeopardy (Xu and Cao 2002). The next year, major flash flooding occurred in almost all major river basins in China, devastating large areas and resulting in the loss of 4,150 lives, the displacement of millions of people and significant damages to property and infrastructure, estimated at about 248 billion yuan, or approximately US$30 billion in 1998 (Sun et al. 2002). Total precipitation in the Yangtze River basin in 1998 was lower and lasted longer, compared to rainfall recorded in 1954, and yet the river experienced record flooding, with eight peaks over two months, indicating a severe reduction in water holding capacity within the basin (Zhang et al. 2000).

Impacts and dependence of the construction and materials sector on forests

In the aftermath of these ecological disasters, it became apparent to both central and regional government bodies that ecological conditions in the upper reaches of the Yangtze and Yellow River basins were affecting the economic welfare and ecological security of millions of people (Zhang et al. 2000). At the time, the Chinese government determined that the removal of 85 per cent of the upper river basin's original tree cover and farming on steep slopes were the primary causes of the drought in 1997, as well as the widespread flooding in the Yangtze River basin in 1998 (Xhao and Shao 2002; Larsen 2002). Intensive logging in areas around the major river systems had led to increased runoff into the rivers, raising river levels as silt was deposited downstream, and thus increased the severity of flooding (Lang 2002). In response, in 1998, the government banned logging in 17 provinces as part of a new Natural Forest Conservation Program (NFCP), intended to run from 1998 to 2010. The main aims of the NFCP were to:

- restore natural forests in ecologically sensitive areas;
- plant forests for soil and water protection;
- increase timber production in forest plantations;
- protect existing natural forests from excessive cutting; and
- maintain multiple-use management of forests (Zhang et al 2000).

The initial investment by the central government in the NFCP, from 1998 to 2000, totalled 22.26 billion yuan, or about US$2.69 billion (Yin et al. 2005). From 2000 to 2010, the State Council allocated a further 96.2 billion Yuan (US$11.63 billion) for forest protection, regeneration, management and relocation of forest workers, and related activities (Yin et al. 2005). Under the NFCP, timber harvests from natural forests in China were reduced from 32 million m^3 in 1997 to 12 million m^3 by 2003 (FAO 2001). The logging restrictions imposed under the NFCP resulted in a significant decline in roundwood production between 1998 and 2003, as shown in Figure 2A.3.

The logging bans and harvest reductions imposed following the 1998 floods displaced a large number of loggers and other forest sector employees, and led state-owned forest enterprises to abandon logging, timber hauling and wood-processing assets worth an estimated 30 billion Yuan (Li 2001; Yin et al. 2005). Moreover, interest payments on loans to these forest enterprises were written off by the government, at a further cost of one billion Yuan annually (Yin et al. 2005). The ensuing reduction in lumber supply also caused timber prices to increase by some 20–30 per cent at the Beijing wood market (Studley 1999).

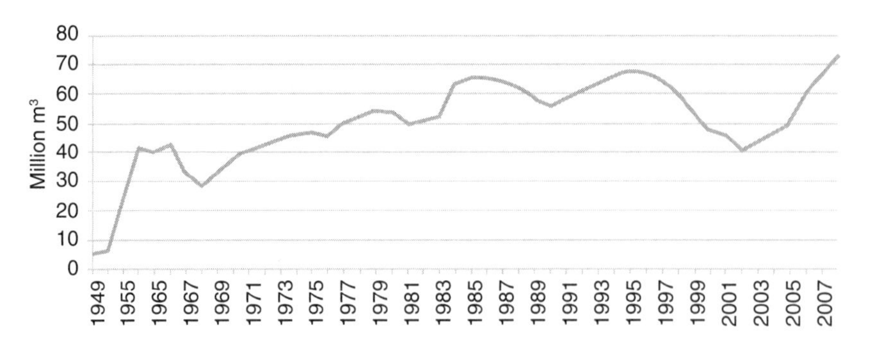

Figure 2A.3 *Production of industrial roundwood in China, 1949–2008*

Source: SFA (2005) for the period 1949–2001; National Bureau of Statistics of China (2009) for the period 2001–8.

Valuing ecosystem losses in timber markets

Over the period 1950–1998, many ecosystem values were not fully taken into account in decisions about China's forest resources, contributing to high rates of deforestation and associated losses of ecosystem services. In a study by Wang Hongchang (1997), the economic impacts of deforestation in China since pre-historical times were estimated by examining the different ecosystems services affected, such as climate regulation, timber and food supply, water regulation, erosion and flood prevention, and nutrient cycling. Based on this study, we estimate here the approximate value of forest ecosystem services lost due to timber production over the period 1950–1998. The study by Wang Hongchang was used for two reasons: first, because it is focused on China; and second, because the estimated value of ecosystem services is directly related to deforestation. A summary of the data used and our analysis is provided in Table 2A.2.

Using data from Wang Hongchang (1997) and other sources, and assuming a linear relationship between deforestation and the value of forest ecosystem services lost, we derive a rough estimate of US$12.2 billion in annual losses over the period 1950–1988. This estimate is then linked to the value of roundwood production over the same period, in order to compare the costs of ecosystem service losses to the market value of timber (McVittie 2010). Specifically, the environmental costs of deforestation are apportioned according to the role of timber production in deforestation (estimated at 60 per cent) and the share of the construction sector in total timber consumption (64 per cent), based on data from FAO (2001) and Zhang et al. (2000). The figure in Box 2.3 shows the results, including the value of ecosystem services lost due to deforestation over the period 1950–1998, as well as flooding damages due to reduced watershed protection and soil conservation services.

Conventional financial analysis considers only products and services that have market prices, and thus excludes forest ecosystems services that are not priced. The observed 20–30 per cent increase in lumber prices in Beijing, following the logging ban in 1998, reflects a small part of the true cost of timber consumption in China. However, based on full cost recovery principles, if the total external cost of ecosystem degradation resulting from timber use by the construction and materials sectors were reflected in market prices, the cost of this important economic input could increase by over 150 per cent.

Table 2A.2 Estimation of forest ecosystem service values in China

Ecosystem service (ES) losses due to deforestation	Valuation method and assumptions	Value of ES lost due to total deforestation from pre-history to 1988: 289.33Mha (in 1992 billion yuan)	Value per hectare (yuan/ha)	Value of ES lost due to deforestation from 1950–98: 83.04Mha (1992 billion yuan)	Value of ES lost due to logging: 59.5% of deforestation (1992 billion yuan)	Value of ES lost due to use of timber in construction and materials: 64% of lumber output (1998 US$/m^3)
Reduced precipitation	Replacement cost: estimated cost of water diversion project (from south to north)	81.00	279.96	23.25	13.83	29.88
Reduced lumber output	Market price: average timber price increase due to scarcity (Fujian) and fuelwood tax (Guizhou)	19.40	67.05	5.57	3.31	7.16
Desertification	Value of production: 50% of crop losses due to desertification are attributed to deforestation	18.80	64.98	5.40	3.21	6.94
Lost water runoff	Replacement cost: additional water 'lost' as runoff due to deforestation valued as per water diversion project	66.70	230.53	19.14	11.39	24.61
Loss of plant nutrients	Replacement cost: value of lost nutrients due to erosion based on the retail price of chemical fertilisers	41.00	141.71	11.77	7.00	15.13
Reservoir and lake sedimentation	Replacement cost: cost of constructing new reservoir capacity to replace sedimentation losses	0.80	2.77	0.23	0.14	0.30

Table 2A.2 Estimation of forest ecosystem service values in China *(Cont'd)*

Ecosystem service (ES) losses due to deforestation	Valuation method and assumptions	Value of ES lost due to total deforestation from pre-history to 1988: 289.33Mha (in 1992 billion yuan)	Value per hectare (yuan/ha)	Value of ES lost due to deforestation from 1950–98: 83.04Mha (1992 billion yuan)	Value of ES lost due to logging: 59.5% of deforestation (1992 billion yuan)	Value of ES lost due to use of timber in construction and materials: 64% of lumber output (1998 US$/m³)
Loss of river transport capacity	Wages: lost income of 1.1 million workers due to 50% reduction in the length of navigable rivers	4.10	14.17	1.18	0.70	1.51
Property loss from flooding	Damage cost: 50% of annual flooding losses are attributed to deforestation	13.40	46.31	3.85	2.29	4.94
One-off flooding damages (1998)	US$30 billion (Yin 1998) divided by cumulative roundwood production over 1950–1998 (SFA 2005)	NA	NA	NA	NA	10.37
Total		**245.20**	**847.48**	**70.37**	**41.87**	**100.82**

Note: For details of the estimation, please contact the author, Mark Trevitt: mltrevitt@gmail.com

China's impact on the global environment

The Chinese economy is increasingly connected with the rest of the world through trade and investment, hence changes in forest policy and forest resource use in China can have significant impacts around the world. As a result of the decline in domestic forest resources and the 1998 logging ban, China's own production has not been able to keep pace with strong domestic demand for timber. The difference is largely made up by imports, which puts increased pressure on forests in other countries.

In other words, the external environmental costs of China's use of timber may have been shifted abroad, raising concerns in some timber-exporting countries, such as Burma, Indonesia and Russia. Over the past decade, unprecedented economic growth in China, coupled with a shortage of domestic resources as a result of deforestation and the logging ban, has turned China into the world's largest importer of unprocessed logs and tropical timber, and the world's second largest importer of wood products. Since 1995, China's imports of wood products have grown by 450 per cent; of every ten tropical trees traded in the world in 2004, five were destined for China (Greenpeace 2006). This raises questions about the strength of forest governance and the sustainability of forest management in countries that are supplying China's demand for timber, even as forests in China itself begin to recover.

Conclusion

The case studies presented above illustrate the importance for businesses to assess the impacts and dependencies of their products and services on ecosystems and biodiversity, throughout their value chains.

In the case of the Aral Sea, the diversion of water to support cotton production and exports pushed the region's hydrologic system beyond the point of sustainability. The agriculture industry effectively kept its production costs down by ignoring (externalizing) the value of environmental damage. If the externalities associated with water use had been included in the costs of cotton production, both the overall output of cotton and the level of irrigation would probably have been reduced. The lesson of the destruction of the Aral Sea is that water resources can and do disappear when used unsustainably, and that changes in ecosystems can have far-reaching impacts. The loss of ecosystem services and the costs of protecting and rehabilitating ecosystems need to be valued and considered explicitly in order to use water resources efficiently.

In the case of China, unsustainable harvesting of timber led to increased scarcity of raw materials, as well as the loss of non-marketed forest ecosystems services. Ultimately, the logging industry lost its licence to operate in many forest areas when the Chinese government imposed a logging ban in 1998. The resulting shortage of timber supplies led to higher costs to the construction sector, narrowing operating margins, disrupting production and potentially increasing market volatility. More importantly, it is clear that deforestation in China over many decades undermined the supply of many valuable ecosystems services, contributing to ecological disasters with significant human and economic consequences.

Each case study serves as a cautionary tale of how the undervaluation of ecosystem services can lead to ecosystem degradation and adverse economic consequences for society and businesses, which are often only recognized after the fact.

References

Brander, Luke (2010) Personal communication, Institute for Environmental Studies, VU University, Amsterdam.

Brander, Luke M., Florax, Raymond J.G.M. and Vermaat, Jan E. (2006) 'The empirics of wetland valuation: a comprehensive summary and a meta-analysis of the literature'. *Environmental and Resource Economics* 33: 223–50 (DOI:10.1007/s10640-005-3104-4).

Cai, Ximing, McKinney, Daene C. and Rosegrant, Mark W. (2003) 'Sustainability analysis for irrigation water management in the Aral Sea region'. *Agricultural Systems* 76(3): 1043–66 (June).

Chapagain, A.K., Hoekstra, A.Y., Savenije, H.H.G. and Gautam, R. (2006) 'The water footprint of cotton consumption: An assessment of the impact of worldwide consumption of cotton products on the water resources in the cotton producing countries'. *Ecological Economics* 60(1): 186–203. (www.waterfootprint.org/Reports/Chapagain_et_al_2006_cotton.pdf).

CNFCM (2000) 'Center for Natural Forest Conservation Management', Unpublished report to the World Bank.

Cohen, David H. and Vertinsky, Ilan (2002) *China's Natural Forest Protection Program (NFPP): Impact on Trade Policies Regarding Wood*, Report prepared for CIDA with the Research Center for Ecological and Environmental Economics, Chinese Academy of Social Sciences.
EJF (2005) *White Gold: The True Cost of Cotton*, Environmental Justice Foundation, London.
FAO (1998) *Time to Save the Aral Sea?* Agriculture and Consumer Protection Department, UN Food and Agriculture Organization, Rome.
FAO (2001) *Forests Out of Bounds: Impacts and Effectiveness of Logging Bans in Natural Forests in Asia-Pacific*, Asia-Pacific Forestry Commission, RAP publication 2001/08, UN Food and Agricultural Organization, Regional Office for Asia and the Pacific, Bangkok, Thailand. URL: www.fao.org/docrep/003/x6967e/x6967e00.htm
Ghermandi, Andrea, van den Bergh, Jeroen C.J.M., Brander, Luke M., de Groot, Henri L.F. and Nunes, Paulo A.L.D. (2009) 'The values of natural and constructed wetlands: A meta-analysis', Tinbergen Institute Discussion Paper TI 2009-080/3, URL: http://ssrn.com/ abstract=1474751
Glazovsky, N.F. (1990) 'Ideas on an escape from the "Aral Crisis". *Soviet Geography* 32(2): 73–89.
Glazovsky, N.F. (1995) *The Aral Sea Basin. Regions at Risk: Comparison of Threatened Environments*, United Nations University Press, Tokyo and New York.
Greenpeace (2006) *Sharing the Blame: Global Consumption and China's Role in Ancient Forest Destruction.*
Kijne, J.W. (2005) *Aral Sea Basin Initiative: Towards a Strategy for Sustainable Irrigated Agriculture with Feasible Investment in Drainage*, Synthesis report, UN Food and Agriculture Organization, Rome (June).
Lang, Graeme (2002) 'Deforestation, Floods, and State Reactions in China and Thailand', Working Paper Series No. 21, City University of Hong Kong.
Larsen, Janet. (2002) *Illegal Logging Threatens Ecological and Economic Stability*, Plan B Updates, Earth Policy Institute (May 21) URL: www.earth-policy.org/index.php?/plan_b_updates/2002/update11 (last accessed 5 February 2010).
Li, Z. (2001) 'Conserving natural forests in China: Historical perspective and strategic measures', Chinese Academy of Social Sciences (working report).
McVittie, Alistair (2010) Personal communication, Land Economy & Environment Research Group, Scottish Agricultural College, Edinburgh.
Micklin, Philip (1988) 'Desiccation of the Aral Sea: a water management disaster in the Soviet Union'. *Science* 241: 1170–5 (2 September).
Micklin, P.P. (1992) 'The Aral crisis: Introduction to the Special Issue'. *Post-Soviet Geography* 33(5): 269–83.
Micklin, P.P. (1993) 'The shrinking Aral Sea'. *Geotimes* 38(4): 14–18.
Micklin, Philip and Aladin, Nikolay V. (2008) 'Reclaiming the Aral Sea'. *Scientific American*, April: 64–71. URL: www.scientificamerican.com/article.cfm?id=reclaiming-the-aral-sea (last accessed 1 April 2010).
Nalwalk, Krilsin. (2000) 'The Aral Sea crisis: The intersection of economic loss and environmental degradation', Graduate School of Public and International Affairs, University of Pittsburgh (April 25).
National Bureau of Statistics of China (2009) *China Statistical Yearbook*, China Statistics Press, Beijing.
Salim, E. and Ullsten, O. (1999) *Our Forests Our Future*, Cambridge University Press, Cambridge, UK.
Severskiy, I., Chervanyov, I., Ponamorenko, Y., Novikova, N.M., Miagkov, S.V., et al. (2005) *Global International Waters Assessment (GIWA) 24*, Aral Sea, University of Kalmar, Sweden.
SFA (2005) *China Forest Resources,* State Forestry Administration, P.R. China, Beijing.
Song, Conghe and Zhang, Yuxing (2010) 'Forest cover in China from 1949 to 2006', chapter 15 in Nagendra, H. and Southworth, J. (eds) *Reforesting Landscapes: Linking Pattern and Process*, Landscape Series 10, Springer.
Spoor, Max, and Krutov, Anatoly (2004) 'The "Power of Water" in a Divided Central Asia', in Mehdi Parvizi Amineh and Henk Houweling (eds) *Central Eurasia in Global Politics: Conflict, Security and Development,* Brill Academic Publishers, Leiden and Boston.
Studley, J. (1999) 'Forests and environmental degradation in Southwest China'. *International Forestry Review* 1(4): 260–5.
Sun, J., Zhao, C. and Wang, L. (2002) *The Long March of Green: The Chronicle of Returning Agricultural Land to Forests in China*, China Modern Economics Press, Beijing, P.R. China.
Temirov, Rustam (2003) 'Lobbying grows in Moscow for Siberia–Uzbekistan water scheme' Eurasianet (February 19).
UNEP (2006) *Challenges to International Waters,* Global International Water Assessment, Regional Assessment 24 – Aral Sea (February).

USDA (2008) *Cotton: Production, Supply and Distribution*, Foreign Agricultural Service, United States Department of Agriculture, Washington, DC.

Wang Hongchang (1997) 'Deforestation and desiccation in China: A preliminary study', in Mao Yu-shi, Ning Datong, Xia Guang, Wang Hongchang and Vaclav Smil (eds) *An Assessment of the Economic Losses Resulting from Various Forms of Environmental Degradation in China*, Occasional Paper of the Project on Environmental Scarcities, State Capacity, and Civil Violence, American Academy of Arts and Sciences and the University of Toronto, Cambridge and Toronto.

World Bank (2003) *Irrigation in Central Asia: Social, Economic and Environmental Considerations*, World Bank, Washington, DC.

Xhao, Guang and Shao, Guofan (2002) 'Logging Restrictions in China: A Turning Point for Forest Sustainability'. *Journal of Forestry* (June).

Xu, J.T. and Cao Y.Y. (2002) 'Converting steep cropland to forest and grassland: Efficiency and prospects of sustainability'. *International Economic Review* (Chinese) 2: 56–60.

Yin, R.S. (1998) 'Forestry and the environment in China: The current situation and strategic choice'. *World Development* 26(12): 2153–67.

Yin, Runsheng, Jintao Xu, Zhou Li and Can Liu (2005) 'China's ecological rehabilitation: unprecedented efforts, dramatic impacts, and requisite policies'. *China Environment Series* 6: 17–32.

Zhang, Peichang et al. (2000) 'China's forest policy for the 21st century'. *Science* 288(5474): 2135–6.

Chapter 3

Measuring and Reporting Biodiversity and Ecosystem Impacts and Dependence

Editors
Cornis van der Lugt (UNEP), Sean Gilbert (GRI),
William Evison (PricewaterhouseCoopers)

Contributing authors
Roger Adams (ACCA), Wim Bartels (KPMG Sustainability), Michael Curran (ETH Zurich),
Jas Ellis (PricewaterhouseCoopers), John Finisdore (WRI), Stefanie Hellweg (ETH Zurich),
Joël Houdet (Orée), Thomas Koellner (Bayreuth University),
Tim Ogier (PricewaterhouseCoopers), Jérôme Payet (SETEMIP-Environnement),
Fulai Sheng (UNEP), James Spurgeon (ERM)

Contents

Key messages

A commitment to proactive management of biodiversity and ecosystem services in business starts with corporate governance and more informed decision making. This implies the integration of BES into business risk and opportunity management, information management and accounting systems. These systems need to support analysis and decision making at multiple levels, including site/project level, group level and product level. BES information is needed for both internal and external reporting. It is important to measure and report on internal processes, but these alone are not sufficient to inform decisions by internal managers or other stakeholders regarding actions and likely consequences.

Businesses can frame BES targets based on principles that define limits such as 'no-go' areas, use a precautionary approach and work towards net positive impact. Business efforts around BES have typically started by identifying what to avoid (e.g. certain activities, technologies or locations). Recently, this has been complemented by emerging concepts and supporting methodologies that define positive aspirations in terms of net impact. Both approaches are valid, given that focusing on 'net' impacts alone may fail to recognize the unique significance of certain natural assets.

There are major barriers to BES measurement and gaps in reporting by business. The economic costs of BES loss are an externality for most companies, which means that they are often not perceived as financially material. For those companies that do report on BES, most treat it in a superficial manner. This is the case even in high-impact industries, and is partly due to limits in guidance available to business on BES reporting, including techniques to translate physical metrics into monetary ones, and partly a consequence of the low priority assigned to BES by reporting organizations.

Measurement of BES in business must expand with support by technical advances. More baseline information on BES is required to support companies in measuring and comparing their own performance. Life cycle assessment (LCA) techniques need to be expanded and refined to enable companies to assess BES along product life cycles and value chains. Environmental management and accounting systems need to capture BES service dependencies and impacts more consistently. Methodological challenges also exist in how information on BES values are incorporated into business planning and decision making systems, rooting it in existing systems rather than adding new parallel systems.

Increased capacity to value BES, together with increased use and continued evolution of existing guidance, will help to improve business accounting and reporting. Improving the ability of companies to value BES through the adaptation of economic valuation tools and changes in the regulatory environment will help establish BES as a more material issue for business accounting and reporting. Existing guidance can be better applied through voluntary efforts and by ensuring that existing management standards provide better support and clarification on BES. A key step is for securities and exchange regulators to provide formal interpretations that can serve as the basis for assessing the materiality of BES with respect to company filings. Further innovation by BES experts in collaboration with the accounting field can help drive standardization, particularly in the area of ecosystem service valuations.

3.1 Introduction

As discussed in the preceding chapter, businesses of all types impact or depend to some degree on biodiversity and ecosystem services (BES) and consequently face a range of risks, but also opportunities. Good business planning requires adequate internal systems to monitor and measure BES to support decision making.

The challenge is to establish reliable information management and accounting systems that can provide relevant information on BES to support operational decisions (e.g. the choice of production technology), to inform financial valuations or project assessments (e.g. capital investment), and for internal and external reporting. The information needs within a company can be diverse and wide-ranging since BES data must be used for activities and decisions at multiple levels: site- or project-level decisions, product decisions and group or corporate decisions. Among the reasons that businesses may use BES indicators include:

- understanding the impacts and dependencies of different business models on BES;
- tracking key performance indicators that relate to strategic business goals and enable effective risk and opportunity management; and
- communicating BES-related performance and challenges to internal and external stakeholders.

The practice of environmental performance measurement in business is well established, but does not address BES as systematically as more 'traditional' areas of environmental management. Within the environmental information and performance tracking systems adopted by many companies, BES represents a special challenge.

Business impacts and dependence on BES are typically more difficult to measure than standard environmental performance indicators, which focus on direct business inputs (e.g. water, energy or materials) and outputs (e.g. pollutant emissions and solid waste).

Effective management of BES requires measurement of business impacts on various components of biodiversity (i.e. genes, species, ecosystems), as well as business dependence on intangible biological processes (e.g. natural pest and disease control, nutrient cycles, decomposition, etc.). In addition, BES assessment requires attention to wider ecological linkages and thresholds, which may lie beyond the boundaries of corporate control. However, existing approaches and tools for environmental measurement and reporting can provide a basis for BES measurement, management and reporting, as well as a foundation for further development of the field.

This chapter looks at the measurement of and reporting on BES impacts and dependencies in business. The chapter first explores the core parameters and goals of BES information systems and then discusses the use of such information in business. The chapter concludes with recommendations to improve measurement, valuation and reporting of BES in business.

3.2 Designing BES information management and accounting systems

This section sets BES accounting in the context of corporate governance and business information systems. It also examines potential BES goals and metrics in general terms.

Designs for stand-alone BES information management and accounting systems, or the integration of BES into existing business information systems, can follow a plan–do–check–act (PDCA) approach that involves an initial assessment stage to define boundaries and materiality, followed by setting objectives and targets, measuring progress on the basis of clear indicators, and backing this up with environmental management and communications systems. The following sections go through these steps, considering what guidance and examples are available at each stage. While much of this involves non-financial data, Section 3.3 will address the role of financial valuation of BES in capital investment decisions.

3.2.1 Corporate governance and accountability: The point of departure

The commitment to address BES systematically starts at the level of corporate governance, the system by which any organization's decisions are made and implemented. Corporate governance is considered here as something that encompasses corporate sustainability and responsibility, going beyond a narrow focus on shareholder value and voting. This implies the consideration of BES impacts and dependencies in relation to overall corporate strategy, together with procedures for their measurement, management and reporting. Biodiversity and ecosystem information systems thus need to be linked to overall business information management, as well as fitting into wider environmental management (see Section 3.2.4).

Integrating BES in organizational strategy requires, among other things, the selection of appropriate indicators for monitoring and evaluation. These may include qualitative indicators, such as a description of the overall management approach to BES issues, or a description of policy and how it is being implemented, as well as quantitative indicators, which may include monetary or physical values.

Much of the effort involved in measuring BES relates to tracking non-financial information that may be important for the company and its stakeholders. This information can also provide a basis for integrating BES into financial valuations. While there are few examples of companies that have published a financial valuation of BES risks and opportunities, many companies have identified ecological systems that merit attention as part of their corporate strategy.

Of course, for some companies, the impacts of BES will be visible, measurable in financial terms and thus considered material. For example, litigation arising from an oil spill and the associated compensation claims for ecological damage may become a significant concern for investors. BES concerns may also influence private investment decisions. For example, Associated British Ports (ABP), Britain's largest port operator, suffered a decline of 10 per cent in its stock market value after the UK government blocked the company's plans for a container terminal at a site in southern England, in April 2004. The plans were rejected largely due to opposition from environmental campaigners, who claimed that the terminal would threaten important wildlife populations (UK Environment Agency 2004).

Even when there is no measurable short-term financial impact, good corporate governance encourages a long-term perspective and consideration of stakeholder relations. For example, the King III Code of Conduct, issued in the Republic of South Africa, states that: 'Governance, strategy and sustainability have become inseparable. It is expected that the company will be directed to be and be seen to be a decent citizen. This involves social, environmental and economic issues – the triple bottom line' (King Committee 2009).

Similarly, in its Guidance on Social Responsibility, the new ISO 26000 standard recognizes the basic principle 'that an organization should respect and consider the interests of its stakeholders'. The four core environmental issues addressed in the new standard include 'protection of the environment, biodiversity and restoration of natural habitats', with further guidance highlighting the importance of valuing, protecting and restoring BES (International Organization for Standardization 2009).

Business's measurement of its impacts and dependence on biodiversity and ecosystems can serve both private and public interests. For example, efforts by business to collect BES data can serve purposes beyond the boundaries of the firm, such as complementing national inventories and state of the environment reports.

3.2.2 Planning boundaries, scope and materiality

In order to track performance in relation to BES, a company must first define whose performance will be measured (the analytical boundary) and what aspects to include (scope and materiality). Company managers may need to consult and agree with experts from other disciplines, such as conservation agencies, on what to measure and over what time period.

A company must also determine from which entities to gather data on BES (e.g. should they consider the impacts of actions by suppliers, subsidiaries, employees or customers?). In the early years of environmental reporting, most organizations measured their impacts by gathering data only from entities over which they had legal ownership and direct control, as required in financial reporting. However, as noted elsewhere in this book, significant aspects of an organization's BES impacts and dependencies may fall outside legal or financial boundaries.

The boundary for BES measurement and reporting can be defined in terms of the intersection of 'significance' and 'control' or 'influence'. In short, BES measurement should focus on the performance of entities that generate significant risks or impacts and over which the reporting organization has control and/or significant influence. Figure 3.1 illustrates these two dimensions in terms of reporting boundaries and the priority entities for monitoring. Business entities in the top right quadrant (i.e. high risk/impact and high level of control) are priorities for BES measurement. Typical examples of significant control or influence include a business subsidiary, where the reporting organization has operational responsibility for a joint venture (regardless of the actual equity ownership), or a buying relationship where the company accounts for a substantial portion of total sales by the supplier.

While the definition of 'control and influence' may be clear, in terms of legal and financial accounting rules, the determination of 'significant impact' may involve more qualitative considerations such as stakeholder perceptions of impacts, alongside scientific analysis of cause and effect relations. Assessment of the significance of impacts on BES requires close collaboration with experts in land, water, biological and physical resources. Note that whether a company has full control of all parts of its external value chain may not matter, as it does not reduce the business risk of a damaging event, which again underlines the need for a consultative approach with suppliers, customers, investors and other stakeholders.

Considerable research has been conducted on the direct and indirect drivers and pressures on BES, as well as the status and trends in BES. The challenge addressed in this book is to define cause and effect relations in terms of business operations. Scientists and managers working in the field of life cycle assessment, using the products of a business as the point of reference (see below), as well as site-level managers looking at their direct and indirect operational impacts, are at the core of

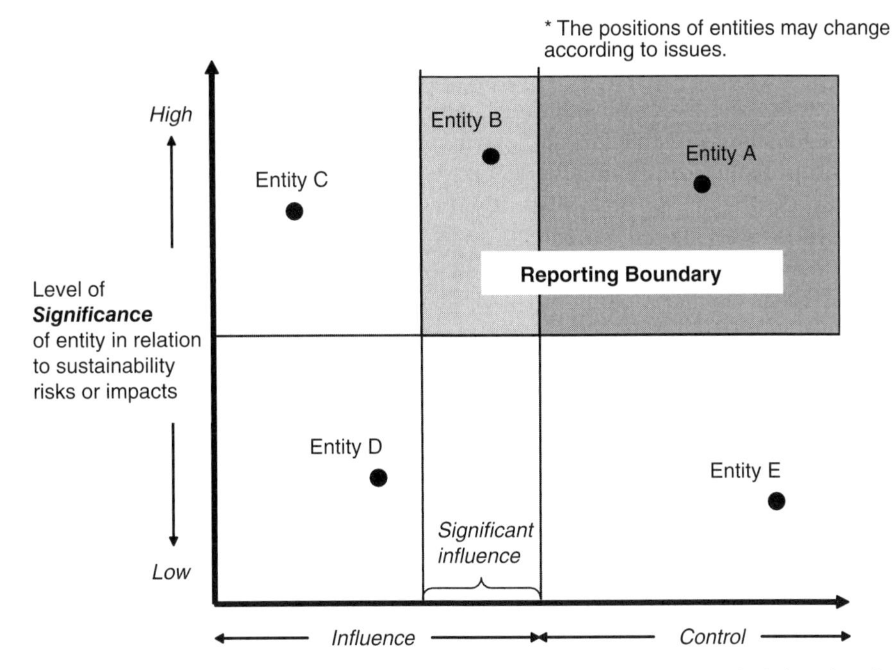

Figure 3.1 *Visual tool to define a reporting boundary*

Source: GRI (2005)

this link. Other points of reference include industrial processes, the production site, business unit, corporate group, supply chain and external value chain. Each of these points of reference has different implications for measurement, the selection of indicators, scope and aggregation of information.

Many companies today set narrow measurement and reporting boundaries that do not reflect key BES issues and entities. Reviews of extractive industry and the food, beverage and tobacco sectors have identified weaknesses in biodiversity targets set by many companies (ISIS Asset Management 2004; Grigg et al. 2009; Foxall et al. 2005). In the food, beverage and tobacco sectors, for example, corporate biodiversity targets generally focused on direct operational impacts rather than indirect impacts within the supply chain.

Companies that manage large areas of land or sea, such as forestry, mining or oil and gas, often find that their own operations represent the most significant portion of their impacts on biodiversity. On the other hand, companies that do not manage land may find that good BES measurement and management requires more extended boundaries. For example, food processors are dependent on the health and productivity of the land of farmers upstream in their supply chains. This highlights again the balance between direct control and influence on the one hand, and level of significance of an issue or entity on the other. Determining the latter can be aided through expert assessment and systematic stakeholder engagement, as set out by the AA1000 standard. Advice on setting reporting boundaries can also be found in the Boundary Protocol of the Global Reporting Initiative (GRI 2005).

Defining operational stages, entities and timing

The concepts of influence, impact and stakeholder interest are reasonable starting points for BES measurement and reporting in business. However, applying these concepts across a large organization can be complex. Such organizations may have multiple points of contact with a range of different ecosystems. For a business with just one major product, service or market, the scope of analysis of their dependence and impacts on BES could be the entire company. For a business with multiple products and services, or active in several markets, the relevant scope may be a particular part of the company. A business may begin with a high-level assessment to identify those parts of the company that have the most significant impacts and/or dependence on BES, followed by a narrower focus for detailed analysis.

Building on methods developed by the World Business Council for Sustainable Development and the World Resources Institute (WBCSD and WRI, respectively; Hanson et al. 2008) for the Ecosystem Services Review (ESR), three basic questions can help managers select an appropriate scope of analysis for biodiversity and ecosystems (Figure 3.2):

1. **Which stage of the value chain?** The starting point for most companies is to examine their own operations, in terms of how impacts on or trends in biodiversity and ecosystem services may affect their business. A useful extension is to look 'upstream' in the value chain, to shed light on how BES impacts and dependence may affect key suppliers, and the business risks and opportunities that these, in turn, may pose to the company. Another approach is to look 'downstream', to gain insight into the implications of BES for the company's major customers.
2. **Who and where, specifically?** When conducting a review of the company itself, certain aspects of the business may be prioritized. Options include – but are not limited to – a particular business unit, product line, facility, project (such as a mine, pipeline, other infrastructure development) or a natural asset owned by the company (such as forest or other landholdings). If the focus is on suppliers, a specific supplier or category of suppliers may be targeted. The scope of analysis may be further narrowed by selecting a particular geographic area in which these suppliers operate. Similarly, in an assessment of major customers, a particular customer or customer segment may be chosen and the scope later refined by selecting a particular region in which these customers are located.
3. **Is the proposed scope strategic, timely and supported?** The scope of analysis should be of strategic importance to the company. Examples include a company's

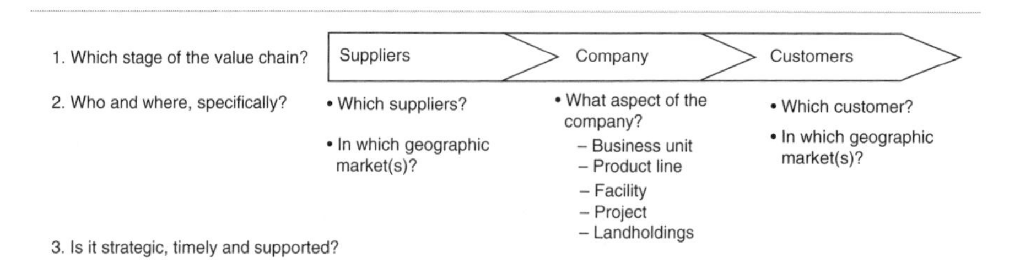

Figure 3.2 *Selecting the scope of BES measurement and reporting*

Source: Hanson et al. (2008)

fastest-growing market, an upcoming major product line or the business unit with the greatest market share and/or profitability. The chosen scope should ideally provide an opportunity to influence pending business decisions. There should be internal support for conducting a review within the selected scope, which implies the need for management buy-in. Experience to date with the ESR shows that it is often most effective to schedule such analysis during regularly planned audits, environmental reviews or strategy sessions (Hanson et al. 2008).

Determining the materiality of BES issues

In addition to defining the boundaries for measuring BES impacts or dependence, a company must also decide which issues to prioritize. Materiality may be assessed in purely financial terms, but this can create blind spots for companies where significant BES externalities exist that are borne by other entities. More nuanced assessments of materiality (or relevance) should consider how actions that may not have an easily measureable financial impact will affect other drivers of business success, such as reputation, licence to operate, employee morale and productivity.

Rather than following the traditional accounting definition of materiality (i.e. an item is material if it could influence decisions taken on the basis of financial statements), some argue that more inclusive definitions should prevail, such as those put forward by AccountAbility (Zadek and Merme 2003) or the Global Reporting Initiative (GRI 2006). This perspective is based on the idea that factors significant enough to substantially concern external stakeholders will ultimately affect the business, particularly if a company looks beyond the very short term. Proponents argue that a strict focus on quantifiable, financial assessments of BES will not properly inform business decision making, because not all relevant risks and opportunities can be translated reliably into impacts on a company's cash flow or financial position.

For those BES issues that are considered material, it is important for a company to articulate, both internally and externally, their relevance from two perspectives:

- whether BES is a material issue for the company as a whole or only for specific operations, regions or products; and
- which aspects of impacts and dependence on BES should be prioritized for action, considering both scale and time frame.

Some work has been undertaken to examine the materiality of BES at sector level. A recent example is work by Oekom Research and Eurosif (2009), which examines both impacts and dependence on BES in agriculture and food, extractive industries, paper and forestry, real estate and infrastructure, and tourism sectors. One of the most comprehensive reports to date is a materiality analysis produced in 2004 by F&C Investments.[1] However, little work has been undertaken to quantify the financial consequences of BES impacts and dependence in business. Nor has the second layer of detail been explored in much detail – namely, what aspects of BES are most material and should be prioritized?

In terms of the ecosystem service categories defined by the Millennium Assessment, the provisioning services of ecosystems represent the most common dependencies and risks for companies. All companies require reliable supplies of inputs, either produced directly or purchased from suppliers as semi-processed goods.

Less obvious but often important are risks associated with impairment of regulating services. These may affect the supply of key business inputs (e.g. climate change may affect the availability of timber, cotton or other agricultural products) or result in negative impacts on other stakeholders that pose a reputational risk or otherwise affect the business licence to operate. Materiality assessments should consider ecosystems in terms of business dependence on underlying ecological processes as well as the direct benefits or services produced by these natural processes.

3.2.3 Principles to consider in setting BES objectives and targets

After identifying boundaries and priorities for BES monitoring and reporting, companies must set clear goals for BES actions. The dilemma for many companies is that almost all business operations will inevitably have some adverse impact on biodiversity and ecosystem services. Furthermore, as noted above, changes in BES do not necessarily follow linear paths, so that it may be unclear whether the actions of a single company, however small, will have negligible impact on biodiversity or if a threshold may be crossed which results in significant ecological change.

Several principles have emerged over time that can help companies define their BES objectives and targets. Simple prohibition (e.g. 'no-go' areas) and the 'precautionary principle' are examples of efforts to express minimum environmental standards or to define clear limits on business operations. The concepts of 'no net loss' or 'net positive impact' have emerged more recently as an aspiration that allows for trade-offs and in-kind compensation for marginal ecological damages, where these can be defined and agreed.

No-go areas

Several companies in the extractive sectors have made voluntary commitments to forgo exploitation of natural resources within certain areas considered to be of high ecological importance or sensitivity, such as tropical rainforests or critical habitat for endangered species (e.g. ICMM 2003; JPMorgan Chase n.d.). Such voluntary commitments can complement mandatory land-use planning to protect sensitive sites and are typically expressed in terms of avoiding areas that have been singled out by international bodies (e.g. World Heritage Sites). To be effective, however, such commitments require universal adherence (i.e. no 'free riding').

Precaution

The causes and consequences of biodiversity loss and ecosystem degradation can be uncertain, and precaution is therefore often advocated in relation to actions that may result in irreversible environmental harm. It is not always possible to obtain clear evidence of a threat to the environment before damage occurs. Precaution – the 'precautionary principle' or 'precautionary approach' – is a response to such uncertainty and has been recognized in both international and national law. The 1992 Rio Declaration, in particular, states that if an action or policy is suspected of causing harm to the public or the environment, in the absence of scientific consensus that such harm would not ensue, the burden of proof falls on those who would advocate taking the action. Part of the justification for precaution is that biological systems are complex and it is often difficult to predict when thresholds or tipping points will be reached. Application of the precautionary principle in the realm of biodiversity is perhaps most obvious with respect to the release of genetically modified organisms (GMOs). The approach is similar to 'no-go' pledges, but typically focuses on the prohibition of technologies rather than avoiding certain geographies.

As with the 'no-go' approach, however, the precautionary principle takes little or no account of economic opportunity costs. Moreover, the effectiveness of the principle depends on its universal application, which may not be realized through purely voluntary action. It also implies that actions will be taken to reduce the scientific uncertainty which prompts its use. Without this commitment, the principle may simply become a tactic for blocking economic activity. When the principle is invoked, plans should therefore be defined to generate the evidence needed to revisit the decision (see Emerton et al. 2005).

No net loss or net positive impact

The concepts of no net loss (NNL), ecological neutrality or net positive impact (NPI) are based on the recognition that many economic activities inevitably result in some residual impairment of BES on a given area of land or sea, even with the best environmental mitigation and rehabilitation efforts. While residual impacts cannot be avoided entirely, a company can aim to achieve a net zero or positive impact by taking actions to reduce threats or rehabilitate BES in other areas, with a view to maintaining overall ecological integrity. Some companies have publicly committed themselves to being net positive or neutral with respect to carbon, water, wetlands, biodiversity or other ecosystems and services. For example, Deutsche Post DHL, Microsoft, Japan Airlines, Marks & Spencer and the Danone Group have committed themselves to climate neutrality across all or part of their operations; Coca Cola has committed itself to water neutrality; Rio Tinto aspires to have a 'net positive impact' on biodiversity; while Sony aims to achieve 'zero environmental footprint throughout the lifecycle' of its products and business activities.[2]

While NNL or NPI can be a powerful aspiration and principle to motivate business action on BES, there are many challenges to achieving such a goal in practice. Some question whether NNL or NPI is technically or politically feasible (see Walker et al. 2009). Nevertheless, the use of biodiversity offsets and related ecological compensation schemes has progressed beyond mere concept and is currently practised extensively across the globe (Madsen et al. 2010). See Chapter 5 for further discussion of biodiversity offsets, in particular.

3.2.4 Measuring and monitoring progress

Most large companies have environmental information systems that gather data from sites and facilities to support local decisions, as well as group-level management. At a corporate or group level, aggregate information on BES may be used by internal and external stakeholders to assess the breadth and depth of the management processes in place and their performance. In general, two broad categories of indicators are used:

- *Process-based*: these measure the extent to which companies have put in place the processes and management systems that, if they are operating effectively, can drive and sustain performance improvements. Examples include the number of sites or facilities that have biodiversity action plans in place, or the extent to which social or environmental impact assessments incorporate biodiversity and ecosystem service impacts and dependence. Such process-based indicators are sometimes criticized on the grounds that they do not provide evidence of outcomes. Moreover, if based only on 'tick the box' completion of procedures, they may suggest progress even when implementation and actual improvements on the ground are minimal.

- *Results-based*: such indicators are intended to provide a picture of performance over time and are essential inputs to any economic valuation of BES impacts and dependence. Results-based indicators tend to be quantitative (e.g. the volume of water abstracted per hectare or tonne of crop produced, or the number of organic product lines in a range). Such indicators are less frequently used than process-based indicators and tend to be customized to individual companies, which can create barriers to benchmarking and interpretation by other stakeholders. Apart from carbon mitigation metrics, which have been the subject of considerable research and standardization, there is currently no consensus on which indicators for BES-related performance may be used across business sectors and regions, due to the variety of different circumstances in which companies operate.

Environmental performance measures to assess the use of natural resources (e.g. energy, water, materials) and the non-product outputs of business (e.g. waste water, air emissions, solid waste, etc.) are relatively well defined within both national environmental regulations and voluntary initiatives, such as the Global Reporting Initiative, ISO14000 series or the Carbon Disclosure Project (CDP). Developing indicators to assess BES performance more broadly is more complex, as it typically involves measuring impacts on or changes to ecological systems, which may extend far beyond a company's operational boundaries or direct control.

Companies can, however, use conventional environmental indicators of resource flows, emissions and pollution as proxies for BES impacts, dependencies and/or responses. For example, the volume and toxicity of waste water discharges may be used as a proximate indicator of potential impacts on biodiversity in receiving water bodies, in the absence of more precise impact data. Conventional environmental indicators may also be relevant when considering how investments in ecosystem conservation or rehabilitation can help improve environmental performance. For example, a company may ask which is more cost-effective – maintaining or restoring the natural filtration and cleaning capacity of a wetland or installing end-of-pipe pollution control equipment.

BES concerns often arise in relation to proposed business investments or other changes in operating procedures at a given time and place, which are easiest to measure and monitor in the context of a specific site or location. The information gathered from individual sites can form the basis for aggregation and decision making across a company or group. For industries with significant direct impacts on terrestrial or marine ecosystems, measuring BES performance at the site or project level forms the basis for BES decisions throughout the organization. At the site level, BES performance data can also be used for managing impacts and developing biodiversity management plans. Different indicators may be needed to address the specific BES challenges and opportunities for a given area and the information requirements of group-level strategies and reporting. For example, a refinery's emissions may have significant impacts on local wetlands, which imply certain monitoring needs. In addition, the same organization may have a group-level goal defined in terms of nutrient loading, which would require additional information.

At the site level, it is often important to set corporate performance in the context of external conditions. Tools such as 'water footprinting', the WBCSD Global Water Tool, as well as the Integrated Biodiversity Assessment Tool (IBAT) can help companies understand the relevance of their impacts on the wider landscape and ecology, and on other stakeholders. Using the IBAT for project planning and site selection processes, for example, allows companies to consider alternative projects, technologies or locations at a point in the decision making process when changes are still economically feasible. Similarly, emerging water accounting or footprinting methodologies enable companies to quantify their operational and supply-chain water footprints, considering: (i) use of blue water (volume of fresh water taken from surface and groundwater resources), (ii) use of green water (volume of fresh water taken from rainwater stored in the upper soil layers as soil moisture), and (iii) cause of grey water (volume of polluted water, usually expressed as the volume of water required to dilute pollutants such that the quality of the receiving water body remains above a certain water quality standard). Box 3.1 provides an illustration of water footprint reporting by SAB Miller, based on the collection of data at site level and the aggregation of results at group level, for a particular product line (litre of beer) and several regions.

Box 3.1 Water reporting by SAB Miller

SAB Miller is one of the world's largest brewers, with operations in over 30 countries. In 2008 SAB Miller announced a commitment to reduce water consumption across its global operations by cutting the amount of water used per hectolitre (hl) of beer to an average of 3.5hl by 2015 – a 25 per cent reduction from 2008 levels. This initiative is designed to save around 20 billion litres of water every year by 2015. The water target is part of a broader strategy which takes a comprehensive, risk-based approach to the company's value chain. SAB Miller reports its average water use per hl of beer, water to beer ratio (hl water/hl beer), regional water to beer ratio and the different sources of water in percentage terms (see charts below). This information is disclosed in the company's annual sustainability report, where it is linked explicitly to both the water strategy and the target to reduce water use per hl by 25 per cent.

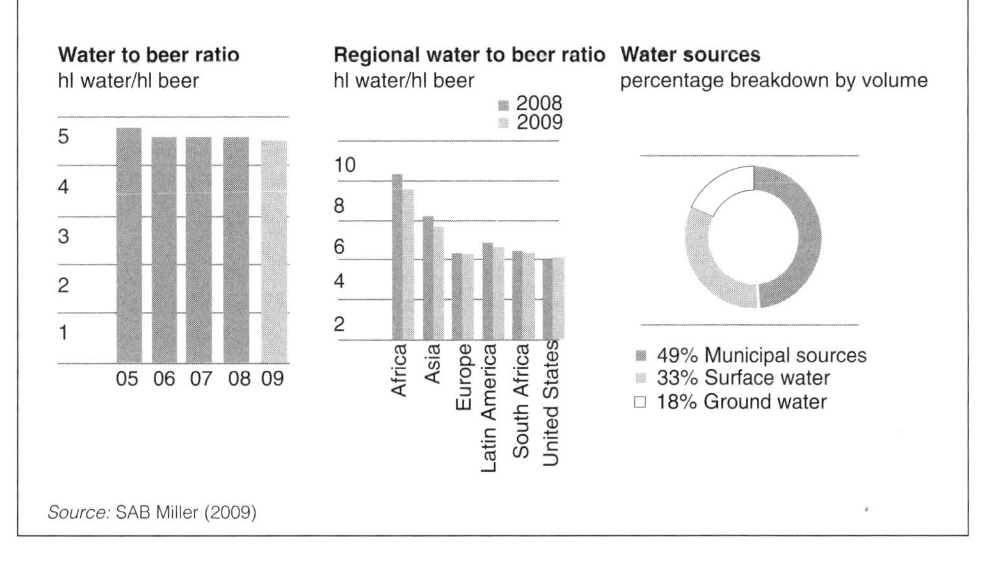

Source: SAB Miller (2009)

3.2.5 Linking BES with mainstream environmental accounting systems

Mainstreaming BES considerations into business planning involves linking general business management accounting with the information systems for environmental management. This does not mean inventing completely new systems and layers of management. Some aspects of BES, such as fresh water use, may already be captured in a company's existing environmental management system. Evidence from public reporting, however, suggests that other aspects of BES are not well integrated into existing environmental management systems, particularly at the group level.

Environmental management systems focus on those environmental interactions that are considered 'significant'. The International Organization for Standardization (ISO) has developed a concept of 'significance' that is similar to the concept of materiality put forward by AccountAbility (2008) and the Global Reporting Initiative (GRI 2006). In the terminology of ISO 14001, environmental management tackles a range of environmental 'aspects' (input and outputs of an activity) and their associated environmental 'impacts'. An aspect is seen as 'significant' if, among other things, it has the potential to cause a demonstrable impact on the environment, and has major financial implications (positive or negative). Definitions of materiality, as noted above, focus on assessing whether the impacts associated with an issue are significant and whether the company's performance in this regard may affect stakeholder decisions. Impacts are often grouped into categories, for example air pollution, water pollution and land contamination.

Management accounting systems provide important information for internal planning, budgeting, control and decision making, and should include relevant BES information. This may include decisions about old or new products, in-house production versus outsourcing, process improvements and pricing. Decisions about products include which materials are required for their production and where these materials can be sourced, with clear implications for BES. Management accounting can incorporate BES metrics and link non-monetary data with financial information. For example, when looking at product manufacturing costs and raw material supply, a company with large impacts on BES may wish to consider potential natural resource scarcities or regulatory restrictions that may affect long-term security of supplies.

In addition, management accounting can provide key inputs for the development of 'balanced scorecard' performance measures (e.g. how business processes may be improved to increase resource efficiency and productivity). Related eco-efficiency targets with implications for BES may include reducing the material intensity of products or increasing the use of renewable resources or certified 'sustainable' inputs.

Environmental management accounting (EMA) has been developed in response to the difficulties of capturing environmental costs within traditional accounting systems. EMA is defined as the identification, collection, analysis and use of information for internal decision making (UNDSD 2001; Savage and Jasch 2005), covering: (a) financial information on environment-related costs, earnings and savings, and (b) physical information on the use, flows and destinies of energy, water and materials (including waste). EMA techniques allow firms to develop and use environmental performance indicators, which may be based solely on physical data or combine monetary and physical data to create eco-efficiency indicators. Physical data may include quantities of ecosystem services used or damages caused to BES, which may or may not be readily translated into monetary values (see Houdet et al. 2009). EMA systems can be aligned with major standards such as ISO14031 or with the indicators specified in reporting initiatives such as the GRI, the Carbon Disclosure Project (CDP) and the Forest Footprint Disclosure Initiative. In current practice, EMA primarily addresses the direct costs of environmental flows by:

- putting a 'price' on non-product output (i.e. pollution and waste), thus highlighting the costs of materials converted into non-marketable waste and emissions; and
- quantifying the monetary impacts of external environmental pressures (e.g. taxes, norms, quotas) in relation to other factors that influence financial results, in order to distinguish transactions of an 'environmental' nature (e.g. compliance costs) from other business transactions.

To provide a more complete picture, companies may supplement EMA by analysing the potential consequences of decisions with respect to intangible environmental assets. While BES involves tangible assets from a public perspective, it is often difficult for businesses to identify and measure their indirect BES impacts and dependencies. This raises the challenge of expanding the scope of BES assessment, going beyond first tier suppliers and clients in the business value chain (this issue is discussed in more detail in Section 3.4).

3.3 Incorporating BES in capital investment decisions

Capital investment is based on identifying viable business opportunities that will generate attractive cash flows within a given time frame. The investment may relate, for example, to entering a new market, expansion, diversification, replacing or upgrading technology. Such decisions are often informed by economic evaluation of alternative options, using 'net present value', 'payback period' or other financial criteria to help business managers decide whether to approve or reject a proposed investment.

The most basic rationale for business investment is to enhance the value of the firm, i.e. creating value for owners. An investor will normally consider a range of value drivers, such as potential growth in sales or taxation, which may be directly or indirectly influenced by 'green' issues.

Biodiversity and ecosystem services may feature at various stages of the investment appraisal process. It could be a decision about entering a market for BES-related goods and services, for example organic food. It could be about diversification into products that use more natural resource inputs. It may involve deciding on expansion into a new country or acquiring a new company in an environmentally sensitive area. It could be about investing in new technology that reduces pollution. It could also be an investment decision with no direct link to BES, but which results in impacts on BES further up or down the value chain.

In considering such options, a business is likely to focus on direct costs and benefits. Making the business case for BES in this context requires the identification of win-win opportunities, based on convincing metrics. While all investment decisions have some BES consequences, these may not translate easily into quantifiable impacts on cash flow, and may therefore have little influence on investment decisions.

Table 3.1 provides an overview of commonly used valuation techniques for business investment, with their implications for BES. Normally the main criterion for approving an investment proposal will be its potential to increase shareholder value

Table 3.1 Commonly used business valuation techniques and implications for BES

Methodology	Key features	Implications for valuing BES
Net present value (NPV) of discounted cash flows	Discounted cash flow (DCF) analysis is the most commonly used investment appraisal methodology in both the public and private sectors. It involves assessing the cash flows that a project, investment or business will generate over a time horizon which encompasses its full life. In order to compare costs and revenues arising at different points in time, future expenses or earnings are normally 'discounted' at a fixed rate, typically based on the investor's weighted average cost of capital. The sum of discounted revenues less discounted costs is known as the 'net present value' (NPV) of the investment, or the value today of the project over its whole life.	Provided all relevant BES impacts and dependencies are accurately valued and included within the scope of business decision making, DCF/NPV offers a plausible framework for investment appraisal. The risk is that certain BES values may be unknown, be mispriced or fall outside the scope of analysis because they do not result in costs or benefits to the investor. The choice of a discount rate can also be problematic, due to uncertainties about the future availability and value of BES.
Internal rate of return (IRR)	IRR is defined as the level of returns which, if used as the discount rate for a particular investment, would result in the discounted costs of the project being equal to discounted revenues (i.e. the IRR is defined as the discount rate which yields an NPV of zero). A higher IRR indicates that the project offers a higher level of return on the initial investment.	IRR can give ambiguous results for projects characterized by negative cash flows at the end their lives. IRR may therefore be unsuitable for projects that involve delayed environmental costs, including remediation expenses incurred at the end of the main operational phase.
Payback period	Payback period is a streamlined investment appraisal technique, employed particularly by SMEs. It is defined as the length of time needed to pay back the initial investment. While the method involves simple calculations, it is necessarily short-sighted both from a business and a BES perspective. A project that is beneficial over the long run is likely to be overlooked if it does not repay the initial investment quickly.	BES impacts that take a long time to manifest would rarely be considered in a payback period calculation. Similarly, other costs incurred at the end of a project's life, such as repairing environmental damage, are also ignored in this evaluation technique, though they are relevant costs to the business.
Indirect valuation	Investors often have to perform an external valuation of companies or other assets. A number of methods may be used to do so, including comparisons of earnings, price to earnings ratios or comparisons to other market benchmarks, such as the value of previous transactions.	Indirect valuation approaches will only reflect BES values to the extent that the target asset or company is properly valued in terms of its impacts and dependence on BES. Use of market benchmarks can be misleading due to the fact that they generally do not reflect BES values.
Informal valuation techniques	Business decisions may be made on 'gut feel', rather than using a formal valuation methodology. The success of this method will depend on the skill and knowledge – and perhaps luck – of the decision maker. Some decisions may seem irrational, but some managers and entrepreneurs have defied conventional wisdom and succeeded in this way.	Informal decision making has both benefits and drawbacks for BES valuation. The approach relies on the investors' personal perspective and values, which may give more or less weight to BES issues depending on the individuals involved.

and profitability. Other criteria, such as protecting the environment, are often secondary as long as legal requirements are met.

There are several barriers to changing 'business as usual' and mainstreaming BES into investment appraisal, notably:

- environmental externalities do not form part of formal business valuations;
- business discount rates often differ from so-called 'social' discount rates;
- businesses may ignore some of the intangible values of BES; and
- there is often limited information and uncertainty about BES values.

These barriers can result in the approval of projects and investments that are less profitable from the perspective of society as a whole than for the owners of the business. Moreover, in some cases, the failure to account for BES impacts or dependencies may undermine the returns to investors, as discussed below.

3.3.1 Barriers to proper valuation of BES in capital investment

Barrier 1: Externalities are missed from business valuation

When applying standard investment appraisal methods, such as those outlined in Table 3.1, and assuming that management seeks to maximize profits and value to the owners, a firm will normally include only the values for costs and revenues that are relevant or material from its own point of view. In other words, a firm may only consider potential damage to an ecosystem in its investment appraisal if it expects to bear some of the costs of the damage itself. Where a firm does not believe that it will bear at least some of the costs of environmental damage, these costs will not normally form part of a formal valuation appraisal. In this case, the cost is an externality, i.e. an impact that is external to the company and its decision making. This also applies to external benefits, such as those arising from ecological restoration activities that do not generate revenue for the investor.

When a business has a detrimental impact on the functioning of an ecosystem, costs may result from reductions in both marketed and non-marketed services. For example, inefficient harvesting by a forest management company imposes a cost on the company since future yields may be reduced. This cost is internal. However, it may also impose non-market costs by reducing the recreational value of the forest. If the company is able to capture some of the value of these services, for example by charging access fees to recreational users, this cost will be partially internalized and may be valued by the business. However, often the costs of the damage will be external. Because many intangible ecosystem services are not valued in the market, they often remain external to investor decision making, particularly if there is no legal basis for external stakeholders to claim for damages, as is commonly the case.

What would change this? Setting aside the personal views of business owners, managers or employees, the fundamental logic for a business is to maximize profits. A business may be concerned about the impacts of its operations on biodiversity and ecosystem services if these are likely to affect the operations of the company itself, or result in reputational damage, delay, litigation or otherwise increase the costs of doing business, reduce sales or create staff recruitment and retention problems. Even when there is not a quantifiable short-term financial impact, changes in BES resulting from investment decisions may still affect the long-term ability of the business to achieve its strategy.

Alternatively, regulation, taxation, subsidies and markets for BES may oblige or encourage a business to consider BES impacts and ensure that ecosystem damage and opportunities form part of its decision making. Environmental regulations could mean that certain industrial processes need to be revised, while taxes and subsidies could alter the pay-offs to investments, and markets would put prices on BES which business would need to buy or sell.

Barrier 2: Business discount rates often differ from social discount rates

Economic theory suggests that an individual will discount future costs and benefits at rates determined by his/her pure time preference and the expected growth in his/her future consumption. In other words, people tend to give less weight to costs and benefits that occur in the future than those arising today, first because they are mortal (and therefore impatient), and second because they expect their income to rise over time. The latter point is subtle but reflects the fact that each additional unit of income delivers slightly less additional utility than the last, due to diminishing returns. If on average people expect to be wealthier in the future, due to general economic growth, we can also expect that an additional increment of income received in the future will generate less utility than the same increment delivered today.

The same logic with respect to future costs and benefits may be applied in business. Because individuals invest in businesses both directly and indirectly (e.g. via pension funds), the discount rates used in business decisions will ultimately reflect the underlying individual discount rates. In addition, businesses expose investors to risk – there is usually no guaranteed return on an equity investment in a firm. So on top of the sources of individual discounting outlined above, business discount rates will include a risk element to compensate investors for the possibility that they may not get their money back.

The standard discount rate used in business investment decisions is the weighted average cost of capital (WACC). The WACC faced by a particular company is established in the debt and equities markets and depends on the preferences and discount rates of the potential pool of investors in that company, as well as the perceived risk of the company or investment project for which capital is solicited. The WACC represents the opportunity cost of investing in the business – forgoing cash that could be spent today or invested in a more secure asset (e.g. government bonds) for the uncertain promise of a higher return later. For most core activities of a business, the WACC is the appropriate rate with which to discount future returns, although for non-core projects a different discount rate may be justified. Normally, a business should only invest in those projects that are expected to create value when discounted at the appropriate WACC.

Typical discount rates seen in the market and generally applied by businesses reflect the savings and investment decisions of institutions and individuals around the world. The longest investment instruments commonly available are 30-year government bonds, which are typically used by pension funds to match their liabilities to provide long-term retirement pensions. While people may and do care about more distant futures, it is currently impossible to derive an appropriate long-term discount rate simply by looking at financial markets. Indeed, for decisions that affect society over the long term, market discount rates are probably not appropriate. Nevertheless, decision makers need a basis to compare present and future costs and benefits. In practice, governments often use a 'social' discount rate for such purposes (Box 3.2). Social discount rates are almost always lower than market rates, reflecting the fact that society as a whole is not mortal (we hope) and is less risk-averse than most individuals.

Box 3.2 Discount rates in UK government planning decisions

The discount rate used to assess UK government policy decisions is the Social Time Preference Rate (STPR). The STPR discounts future consumption, not future utility. The STPR is the sum of two elements:

- the pure rate of time preference, reflecting the fact that individuals discount future consumption, irrespective of changes in per capita consumption; and
- a component reflecting the fact that, if consumption is growing over time, an additional unit of consumption in the future will bring less additional utility than an additional unit of consumption today (reflecting diminishing returns).

As an illustration, the UK government uses an annual STPR of 3.5 per cent for its planning decisions (HMT Green Book), based on a discount rate of 1.5 per cent (accounting for risk of death and pure time preference) plus an estimate of long-run per capita income growth of 2 per cent annually. The STPR reflects the assumption that each increment of future consumption would be worth half as much to people who are twice as wealthy as the current generation (i.e. the marginal utility of consumption is assumed to have an elasticity of one). The STPR is used to evaluate public investment decisions up to 30 years' duration. Lower discount rates are used for longer time horizons, due to uncertainty about the distant future.

Source: UK Government, HM Treasury www.hm-treasury.gov.uk/data_greenbook_index.htm

Business decisions that affect biodiversity and ecosystems are often made using relatively high, market discount rates. The danger is that, by doing so, business may be undervaluing potential adverse impacts on future generations. In addition, even the social discount rates commonly used by governments may be too high. As set out in Box 3.2, it is assumed that consumption growth will be positive in the future. However, biodiversity loss and ecosystem degradation is likely to result in lower levels of consumption of some ecosystem services in the future. Some of this damage may not be reversible at any cost, and there is a limit to which other resources or technologies can substitute for the loss of ecosystem services. This implies that future generations may be worse off than our own, at least with respect to the supply of ecosystem services, in which case a lower or even negative consumption discount rate may be justified. Further discussion of discounting and its application to BES, as well as related ethical issues, can be found in TEEB Foundations (2010: chapter 6). This companion book concludes that a variety of discount rates, including zero and negative rates, may be appropriate, depending on the time period involved, the degree of uncertainty and risk, ethical considerations around intra and intergenerational equity, as well as the scope of the project or policy under consideration.

Barrier 3: Businesses may not account for intangible values

It is well known that biodiversity and ecosystems generate both tangible and intangible values. The latter include so-called 'non-use' values, defined as the value that people place on an ecosystem or resource that is not related to any direct or indirect use, including the value of species and habitats for spiritual, aesthetic, heritage or bequest reasons. Some argue further that ecosystems or component species have an 'intrinsic' or moral value, independent of human preferences (see TEEB Foundations 2010: chapter 4). In addition, ecosystems provide regulating, provisioning and supporting services which can be important even when they are difficult to measure or value.

These and other intangible benefits of BES are not generally traded in markets and their value is not widely understood or agreed. Such values may be reflected in public policy, for example through environmental restrictions on certain activities in certain locations, but there remains wide disagreement about the importance of intangible benefits, which makes it difficult to include loss or gains in such services in business investment decisions. As noted above, firms may not bear the costs of damaging (or reap the rewards of providing) intrinsic or intangible values, except where:

- regulation limits damage to ecosystems or requires remediation or compensation (e.g. restrictions on pollution, production capacity or total output, restoration or offset requirements);
- environmental damage affects the firm's own revenues, now or in the future (e.g. changes in real estate value due to loss of environmental amenity);
- payments or other incentives are provided by other firms, NGOs or public agencies; or
- adverse publicity damages the firm's reputation or brand.

For these reasons, businesses may have an interest in valuing their impacts on biodiversity and ecosystem services, including intangible values. This is increasingly possible using tools developed by economists to measure the non-use values of environmental assets, including so-called 'existence' values. The contingent valuation method, for example, asks individuals about their own valuations of intangible environmental benefits and is the main method used to assess non-use values (Box 3.3). Contingent valuation has gained wider acceptance since its use to value damages resulting from the Exxon-Valdez oil spill (Carson et al. 1992). After an 18-year legal battle, the US Supreme Court finalized Exxon's compensatory payments at over US$500 million, including damage to so-called existence and other non-use values of biodiversity.

Barrier 4: Limited information and uncertainty

Businesses may find it difficult to value BES risks and opportunities accurately, due to lack of scientific and economic data or regulatory uncertainty. Although financial modelling and other valuation techniques can account for risk, they require that a specific probability is placed on every eventuality. The difficulty in assigning probabilities to potential BES outcomes is a major barrier to including them in business valuations. This is exacerbated by the absence of standard metrics with which to monitor BES impacts and dependence.

Box 3.3 Contingent valuation in the Kakadu Conservation Zone

In the early 1990s, the Resource Assessment Commission (RAC) set up by the Commonwealth of Australia investigated various options for the use of natural resources within the Kakadu Conservation Zone (KCZ). Alternatives included opening the KCZ for mining, or combining the KCZ with the adjoining Kakadu National Park (KNP). The KCZ is believed to contain significant reserves of gold, platinum and palladium. Environmental groups argued that potential damage from mining was likely to extend beyond the KCZ to the KNP, and that this was significantly detrimental to the public use and non-use values of the park. Conversely, the mining company sponsoring the proposals argued that damage would be minimal and that the public did not place a high value on the KCZ.

The RAC's investigation involved two major components. First, a study was undertaken to estimate the likely risk of damage from mining. Second, the RAC used a contingent valuation (CV) survey to estimate the economic value of potential damages. Because the extent of damage was unknown when the survey was undertaken, major and minor damage scenarios were considered. Based on a description of the KCZ and the potential environmental damage scenarios, respondents across Australia were asked if they would pay a predetermined price to avoid the damage. By randomizing the prices proposed to each respondent, an average willingness to pay (WTP) could be estimated while controlling for differences in the characteristics of the sample population.

The results of the study implied that public WTP to avoid damage to the KCZ, at AU$435 million, far exceeded the net present value of the proposed mine, estimated at AU$102 million. The total value of avoiding mining damage was obtained by multiplying the median WTP to avoid the minor impact scenario (AU$80 per household surveyed) by the total number of households living in Australia. Following the RAC's report, the Australian government decided not to issue a permit to mine the KCZ site. Interestingly, the results of the CV study were not included in the final RAC report, perhaps due to uncertainty (at that time) about the validity of non-market valuation methods. Nevertheless, this example demonstrates the potential of economic valuation techniques for assessing the value of ecosystem services in a project appraisal setting, and highlights the fact that intangible values can be measured to some degree. Such an approach can help firms to establish the potential costs of damages associated with their investments. Project planners can also use such techniques to identify configurations and methods that would have the least impact on intangible ecosystem values.

Source: Carson et al. (1994); Imber et al. (1991)

Another characteristic of BES decline is that it can be hard to predict or subject to sudden, unexpected change. A small amount of degradation may have little effect on the services that people obtain from ecosystems. However, as the level of harm increases, ecosystem services may be lost at an accelerating rate. Thresholds or 'tipping points' may also arise, beyond which an ecosystem enters a new state and the supply of certain ecosystem services is significantly reduced. Moreover, in some cases, ecosystem service losses may be irreversible on a human timescale.

Finally, companies are taking actions today that will be affected by changing norms and regulations in the future. Costs may materialize over the life of a project or investment, due to new regulation, but will only be considered if business managers think they will occur or are likely to occur. As with other risks, regulatory impacts can be included with a discrete probability if this is known. However, as seen in the case of climate liabilities, regulatory uncertainty may simply lead firms to ignore the future costs of their actions. Importantly, this is likely to be more common for environmental opportunities than risks – an investment in ecosystem assets such as biodiversity credits may be viewed as 'speculative', while acknowledging that a firm undertaking a damaging activity may need to pay could be considered 'prudent'.

3.4 Collecting and using information at the product level

Life cycle management provides a practical approach for product-based decision making in business, which may incorporate BES aspects. Life cycle management typically combines product-level assessment tools, such as life cycle assessment (LCA), with environmental management (such as ISO 14001) and reporting systems. Life cycle management looks beyond a particular industrial site or stage of the value chain to assess the full impact – including socio-economic impacts – associated with a product or service throughout its life cycle. This section reviews recent efforts to integrate BES information in LCA methods.

3.4.1 A brief overview of life cycle assessment

Life cycle assessment (LCA) is used to study the environmental interventions and potential impacts that arise throughout a product's life, from raw material acquisition through production, to use and final disposal (i.e. from cradle to grave). For instance, the life cycle of a tomato would include the production of fertilizer, pesticides, water, peat or other media for seedling production, energy for heating greenhouses and transport, packaging, processing energy (e.g. cooking) and waste treatment.

The aim of LCA is to provide information to enable business to reduce resource consumption, emissions and other environmental impacts at all stages of a product's life. LCA can be used to compare different products (e.g. bio-fuels with fossil fuels) or to identify environmental issues and thus potential improvements along the life cycle.

Standard phases with related questions in the LCA are set out in Figure 3.3. The life cycle inventory (LCI) involves data collection and calculation procedures to quantify relevant inputs (resources) and outputs (emissions). The life-cycle impact assessment (LCIA) aims to analyse and evaluate the magnitude and significance of the potential environmental impacts of a product system. LCIA is needed because in the inventory analysis several hundred different emissions and resource uses may be quantified, and a comparison of two products or scenarios based on so many environmental interventions is virtually impossible. In the LCIA, an aggregation of these environmental interventions according to the type of impact or damage is performed. This reduces the number of environmental indicators to between 1 (for fully aggregating methods) and approximately 10, which is far easier than comparing hundreds of emissions and resource use flows.

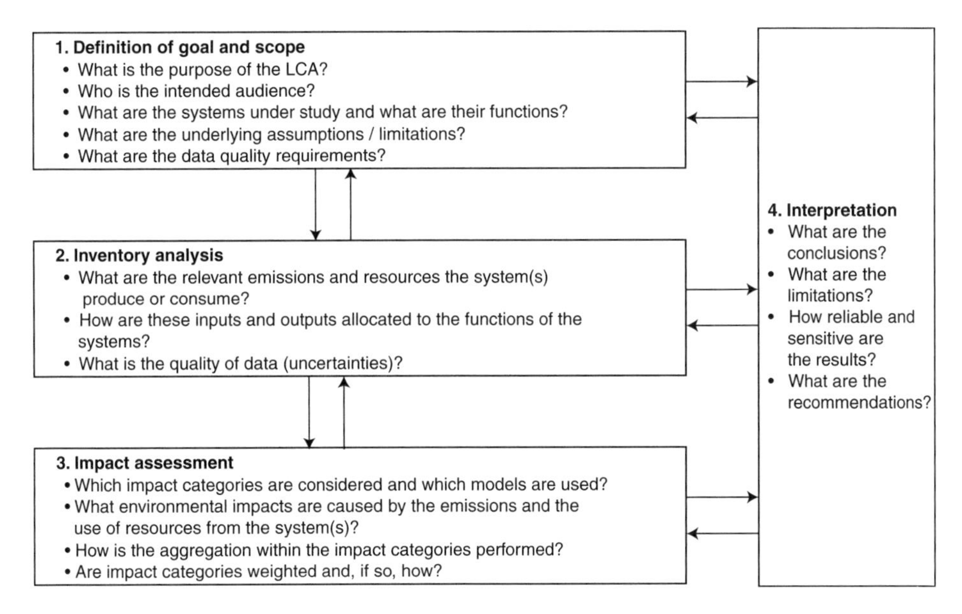

Figure 3.3 *Four phases of life cycle assessment*

Source: ISO standards 14040 and 14044

The overall framework for life cycle assessment is described by de Haes et al. (1999) and elaborated in the International Reference Life Cycle Data System (ILCD) Handbook (European Commission 2010). This framework defines the relationship between environmental interventions (i.e. modification of the environment arising directly from a business activity) and the resulting impacts. Due to the complexity of the causal chain, in many cases, several steps are considered between alterations of the environment and final impact (also called 'endpoint'). Each intermediate point of measurement along the causal chain is called a midpoint (such as ecotoxicity, eutrophication, land use, etc.). Beyond this, the endpoint refers to an 'area of protection', which means entities of ultimate interest to society, such as human health or the quality of the natural environment.

Over the last 20 years, several LCIA methods have been developed. These differ in terms of their definition of impact categories, consideration of environmental compartments, number of emissions and resources considered, and level of aggregation. There are methods that stay at the level of impact categories and others that aggregate further to the level of damages to the three areas of protection: natural environment (ecosystem health), human health and resources (see Figure 3.4). Different weighting techniques can also be used (e.g. based on targets set by government or by experts).

3.4.2 Integrating BES in LCA

Biodiversity-related endpoints are not currently well integrated in LCA methods and guidelines. Various approaches are being examined under the UNEP/SETAC Life Cycle Initiative (http://lcinitiative.unep.fr). Some approaches estimate a potential

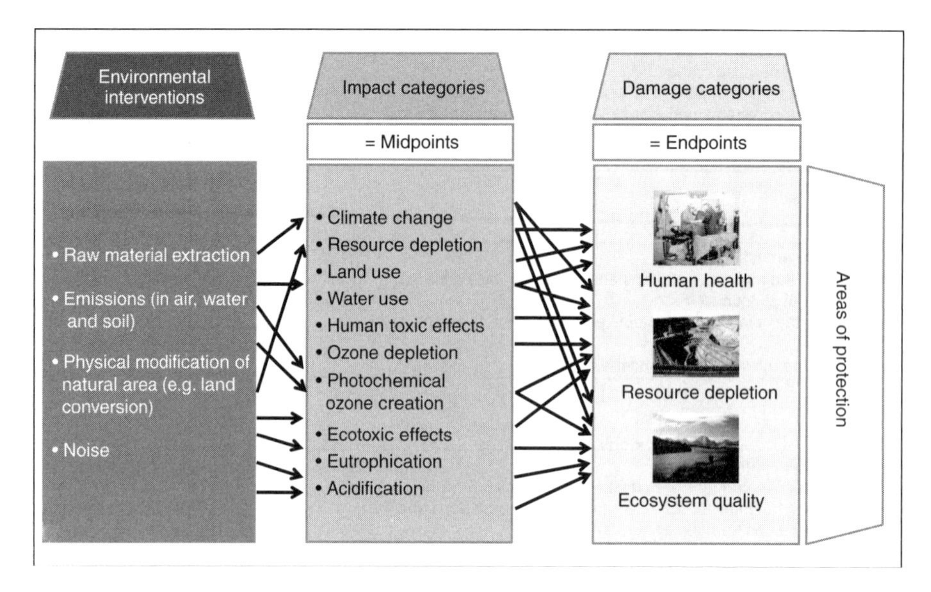

Figure 3.4 *Assignment of interventions, impact categories and damage categories*

Source: Jolliet et al. (2003, 2004)

percentage change in ecosystem diversity, independent of the location and time of the impact. Other emerging methodologies express damage in terms of the fraction of species eliminated over an area and time period.

LCA incorporating BES needs to account for a wide range of impacts, such as ecotoxicity or land-use change. These impacts result from different environmental interventions. In the case of ecotoxicity, for example, the impact is due to emissions of toxic substances into the environment. Thus, impact is related to the quantity of a particular substance released into the environment and is proportional to the hazard or toxicity of the substance. Ecotoxicity may be assessed by modelling exposure concentrations and examining species level indicators of abundance or reproductive success, and extrapolating as appropriate from one or more indicator species to assess impacts on entire ecosystems.

Land-use impacts are proportional to the surface area transformed, the extent or magnitude of transformation, and the ecological sensitivity or value of the area. Measurement of land transformation must take account of the type of land use before transformation, the land use after transformation, the geographic extent and a 'relaxation' period. The LCI of a product system thus shows different land-use types, and gives information about their quantities in space and time. In the LCIA, these are weighted with respect to their potential ecological value or impact.

Land use in particular influences biodiversity through habitat change, fragmentation and pollution linked to intensive agriculture, forestry and the expansion of urban areas and infrastructure. The measurement of land-use impacts on biodiversity, however, is a complex task. The UNEP/SETAC Land Use Working Group distinguishes between biodiversity damage potential (BDP) and ecosystem services damage potential (ESDP).

BDP addresses the 'intrinsic' or conservation value of biodiversity. It is based on factors for different land-use types and intensity classes (see Koellner 2003, and Koellner and Scholz 2008). BDP accounts for the diversity of plants and explicitly considers threatened species. This diversity is then related to regional mean species abundance as a reference or benchmark. Until now, most of the quantification has been carried out for European land use. However, Schmidt (2008) compared the impacts related to the occupation of one hectare per year in Denmark, Malaysia and Indonesia. Such a globally applicable method needs further development in order to assess global resource flows and associated land-use changes.

While LCA is increasingly applied to product choices and optimizations, it has some limitations. For many impacts, in particular those related to BES, their magnitude depends heavily on spatial conditions. For instance, converting tropical rainforest into a bio-fuel plantation has a different impact on biodiversity than producing bio-fuel crops on land previously used to grow food or other commodities. Additionally, the main impact often occurs in locations far away from the place of final consumption. This is especially the case for agricultural commodities, which cause a variety of ecosystem impacts at the place of cultivation and are then often exported.

With increasing globalization, product supply chains have become more complex and difficult to track. Nevertheless, the growing interest and perceived responsibility of consumers and companies in importing countries, as well as the increasing loss of BES in producing countries, implies a need for better information along the product value chain.

One major strength of LCA is that it considers the whole life cycle. LCA practitioners are developing tools that allow for greater spatial differentiation. However, it may be difficult, even impossible, to track the entire value chain of all products with a high degree of spatial resolution. LCA is thus not able to provide a complete substitute for site-level assessment. At the same time, it should be noted that many companies have an influence on the entire life cycle of a product, in one way or another. LCA can help companies use this influence more effectively.

3.5 Collecting and using information at the group level

The combination of dependencies, impacts, risk and opportunities associated with BES at the site and product levels together constitute the overall BES profile for a company. Box 3.4 provides an example from the field of carbon measurement and reporting, based on typical company data but 'anonymized' to respect confidentiality. This box shows how environmental indicators collected at product and activity level can be aggregated at group level.

The tools used to evaluate a company at group level, in terms of BES, are often very different from those used at the site or product level. At group level, a company could be expected to:

- have relevant BES policies and procedures in place;
- monitor performance relative to those policies;
- consider BES in financial analysis and decisions; and
- publicly report on their relationship to BES.

Box 3.4 Carbon reporting by Typico plc

Developments in carbon reporting provide a benchmark for the progress that may be expected in reporting on biodiversity and ecosystem services in the coming years. In 2009, an example of carbon and climate change reporting for a fictional technology company 'Typico plc' was prepared by PricewaterhouseCoopers and is illustrated below.

Key indicators	Direct company impacts		
Climate indicators	Financial performance		
Emissions	2007 (£m)	2008 (£m)	2009 (£m)
• Net costs/(savings) from GHG reduction activities	10.2	15.1	(3.5)
• Net costs/(savings) from 'eco products'	2.3	0.7	(2.4)
• Carbon offset expenditure	6.5	7.8	9.8
Net costs	19.0	23.6	3.9

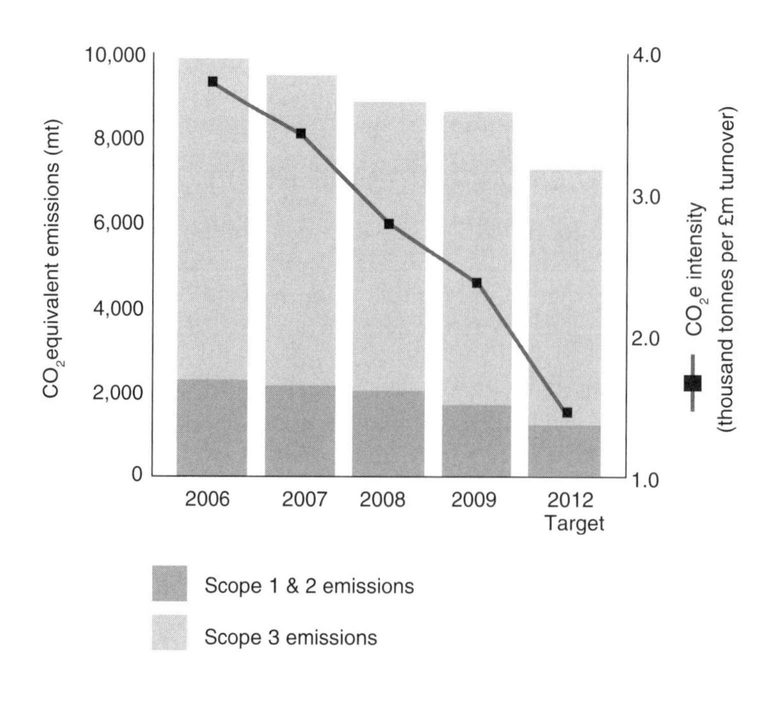

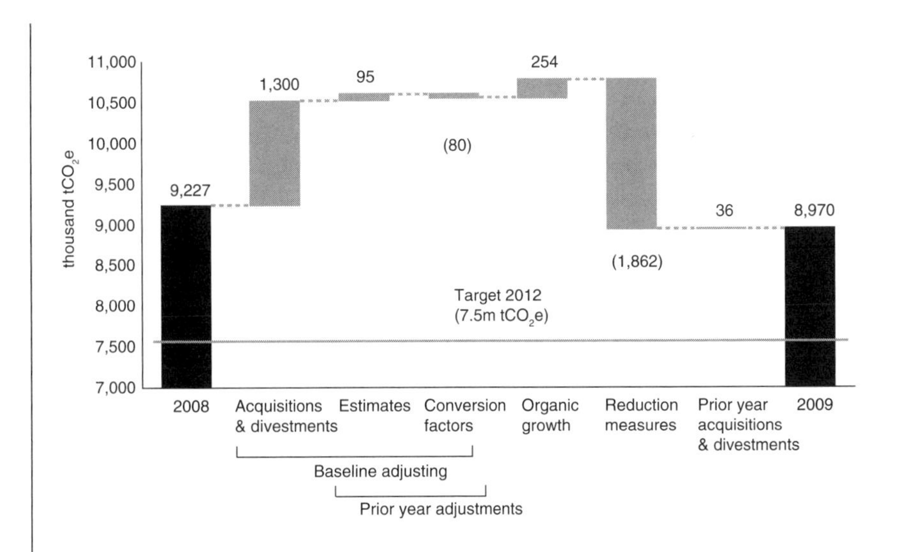

This example was included in the exposure draft of the Climate Disclosure Standards Board's (CDSB) reporting framework (www.cdsb-global.org/). As of 2010, some 3,000 organizations in 60 countries measured and disclosed their greenhouse gas emissions, water management and climate change strategies through the Carbon Disclosure Project (CDP). The CDP is among the most widely adopted voluntary environmental disclosure initiatives and acts as the Secretariat to the CDSB. A recent analysis of participation in the CDP found that US firms were more likely to comply with voluntary climate reporting requirements if they (or other firms in their industry) were the focus of shareholder environmental resolutions or operating in states where more stringent climate regulations were anticipated (Reid and Toffel 2009).

Sources: PricewaterhouseCoopers for TEEB Foundations (2010); Reid and Toffel (2009)

Tools are available to assist companies in these practices, although working at a group level bring its own challenges. For instance, it is generally not possible to generate aggregate BES numbers at group level, simply by adding up site-level or product-level impacts, due to the heterogeneous nature of BES impacts. Moreover, existing methodologies and tools, particularly those related to economic valuation, provide limited guidance or emphasis on BES. Hence this section focuses on the integration of BES impacts in group-level financial accounting and public (financial and sustainability) reporting.

3.5.1 Financial accounting standards and BES

Financial accounting and reporting differs from life cycle accounting, environmental performance measurement and other types of management accounting in that it is primarily intended to serve external audiences rather than internal users. Over the last decade, and especially in the past two years, there has been widespread debate over the purpose of financial reporting. According to the International Accounting Standards Board (2001), the objective of financial statements is 'to provide information about the financial position, performance and changes in financial position of an entity that is useful to a wide range of users in making economic decisions'. They go on to say that:

> Financial statements prepared for this purpose meet the common needs of most users. However, financial statements do not provide all the information that users may need to make economic decisions since they largely portray the financial effects of past events and do not necessarily provide non-financial information. (International Accounting Standards Board 2001: 12–14)

Defining the purpose of financial reporting in terms of the needs of a narrow class of stakeholders – investors and lenders – influences the extent to which such reporting can address issues such as BES. This is because the criteria developed to ensure relevant and reliable financial reporting for the purposes highlighted above are almost inevitably framed in such a way as to exclude so-called 'intangible' issues, such as BES impacts or dependencies.

At the heart of this disconnect is the accounting concept of 'recognition'. This suggests that for an item to be recognized as an asset or a liability by an entity, it must be considered probable that any future economic benefit associated with the item will flow to or from the entity and that the item has a cost or value that can be measured reliably. For accounting purposes, an asset is a resource controlled by an entity as a result of past events from which future economic benefits are expected to flow to the entity, and a liability is a present obligation of an entity arising from past events, the settlement of which is expected to result in an outflow from the entity of resources embodying economic benefits.

The vast majority of ecosystem services and the vast bulk of biodiversity fall outside these criteria and are thus neither accounted for internally by organizations (in the public or private sectors), nor are they (or managers' stewardship of them) reported externally in standard financial statements. The main exceptions to this rule occur where:

(a) A recognizable market exists which gives rise to 'reliable' valuations. Examples include agricultural land and produce, forestry, fish-farming or carbon trading in some jurisdictions. For companies operating in relevant sectors, standard accounting valuation rules are applied to stocks of land, timber, crops, livestock or other 'inventory' items, in order to price transactions or value assets and liabilities.
(b) An enterprise operates in a sector where stewardship of BES is fundamental to its licence to operate. An example is the UK National Forest Company (Box 3.5), whose annual reports and accounts contain a wealth of information dealing with the management's stewardship of the natural resources in the charge of the enterprise.
(c) The organization is located in the public or not-for-profit sectors and is subject to (or volunteers for) detailed accounting of BES assets and liabilities. This is similar to the National Forest Company example, but the general purpose of the organization is to provide services that support the public good (e.g. local authorities or government departments). It should be noted, however, that most public agencies and NGOs nevertheless do not account for BES assets and liabilities.

Box 3.5 UK National Forest Company: 2008/2009 Annual Report and accounts

Objectives:

- To secure further forest creation, contributing to the delivery of targets contained within the Forest Strategy 2004–2014 and the National Forest Biodiversity Action Plan.
- To achieve a high-quality, sustainable National Forest.
- To demonstrate a leadership role in responding to climate change, both in forest creation practice and in work with other forestry organizations to develop a national approach to domestic forestry and climate.
- To realize the economic potential of the forest, building on its environmental foundations, and to consolidate the sustainable development achieved to date.
- To make further improvements to access and participation in the Forest, broadening the range of people using and enjoying it.

Headline achievements – significant activity in the year included:

- The National Forest recognized as one of the first three UK examples of the European Landscape Convention.
- The Woodland Owners' Club re-established, as a driver for good woodland management, including dealing with pests and diseases.
- A comprehensive bird survey completed, applying national surveying methods.
- Further high-quality visitor signage and furniture installed in the East and West Midlands. (The tourism economy was confirmed as being worth more than £260m a year.)

Source: UK National Forest Company (www.nationalforest.org)

While BES typically falls outside traditional business accounts and reporting, the World Business Council for Sustainable Development (WBCSD), working with several public and private entities, has developed guidelines for business on how corporate accounting practices can incorporate BES values more explicitly. The *Guide to Corporate Ecosystem Valuation* (WBCSD 2011) identifies several reasons why business may wish to value ecosystem services, including:

1. *Improving decision making*: ecosystem valuation can be used to strengthen internal management planning and decision making related to environmental impacts or the use of natural resources.
2. *Informing mindsets*: ecosystem valuation can help raise environmental awareness and support for business action among employees, shareholders and consumers.
3. *Sustaining and enhancing revenues*: ecosystem valuation can help companies assess the benefits of participating in markets for green products and services, by determining whether the returns are sufficient to warrant investment.
4. *Reducing costs and taxes*: ecosystem valuation may be used to identify potential cost savings from maintaining or creating ecosystems. Companies may also be

eligible for reduced taxes for managing assets to generate ecosystem services which yield broader social benefits.

5. *Revaluing assets*: ecosystem valuation allows companies to quantify the values of natural assets that they own or have access to, through determining the broader benefits they provide and identifying ways to capture that value.
6. *Assessing liability and compensation*: as environmental regulation becomes more stringent, companies face an increasing array of penalties, fines and compensation claims when their operations damage ecosystems. Ecosystem valuation can help to inform project appraisal and risk assessment to minimize such threats, and gauge the cost of ecosystem damage if claims are made against companies.
7. *Measuring company value*: ecosystem valuation provides a means to quantify environmental performance improvement and allow external sources to factor this in their evaluation of company and theoretical share values.
8. *Reporting performance*: ecosystem valuation can help measure a company's environmental performance, and facilitate more complete reporting and disclosure. Corporate ecosystem valuation (CEV) can also be used to indicate the value of externalities or to form the basis of case studies outlining how a company may lead the way in accounting for wider environmental and social impacts.
9. *Optimizing societal benefits*: ecosystem valuation can help inform stakeholder negotiations and strengthen the decision making process by facilitating better coordination and planning with other stakeholders. It can also be used to help select alternatives that provide a net positive or maximum benefit to society.

In reality, the idea that biodiversity or ecosystem services have economic value is scarcely reflected in the measures conventionally used to assess and report on company performance, or to weigh alternative business opportunities and risks. As a result, business decisions are made based on a partial understanding of costs and benefits. However, the ability to factor BES values into corporate decision making is becoming more important, as new markets for ecosystem services are developed and as new regulations increasingly require companies to measure, manage and report their BES impacts.

The extent to which economic value (let alone market prices) can or should be the basis for decision making about BES is open to debate. We recognize the limitations of economic valuation but also argue that information about ecosystem values is generally helpful and rarely harmful (see also TEEB Foundations 2010: chapter 4). Given the current limitations of markets and accounting, BES values rarely appear as a material component of financial accounts. More generally, as noted above, few companies apply economic valuation to their BES impacts and dependencies.

3.5.2 Public reporting

While biodiversity and ecosystem service values can be important to companies, as illustrated above, the issue is generally not well reflected in public reporting by business. Analysis by PricewaterhouseCoopers shows that of the 100 largest companies in the world in 2008 by revenue, only 18 made any mention of biodiversity or ecosystems in their annual reports. Of these 18, just 6 companies reported measures to reduce their impacts and only 2 companies identified biodiversity as a key strategic issue. Of the same 100 companies, 89 publish a sustainability report. Of these, 24 disclose some measures taken to reduce impacts on BES, while 9 companies identified impacts on biodiversity as a key sustainability issue (Figure 3.5).

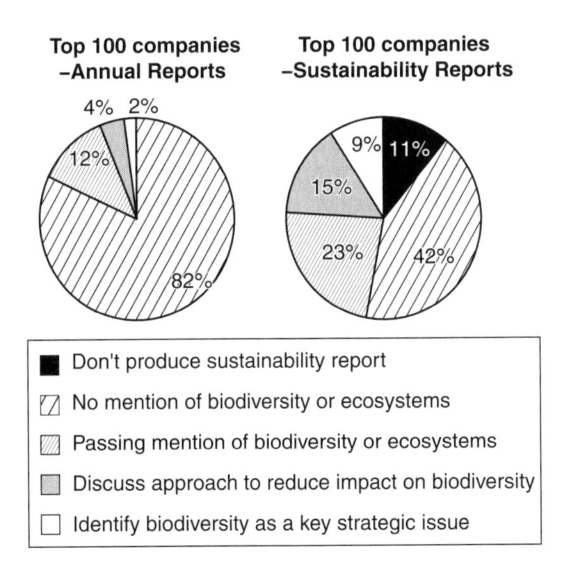

Figure 3.5 *Reporting on biodiversity by the 100 largest companies in 2008*

Source: PricewaterhouseCoopers for TEEB

Further analysis focused on a subset of the 100 largest companies, including only those that were identified as falling into either high biodiversity impact or high biodiversity dependent sectors. Looking first at these 36 companies' annual reports, we observe a similar pattern as for the top 100, with almost the same proportion of biodiversity reporters in each category (Figure 3.6). Looking at these companies' sustainability reports, however, we observe a significantly larger proportion of the high dependence or high impact group that identify biodiversity as a key strategic

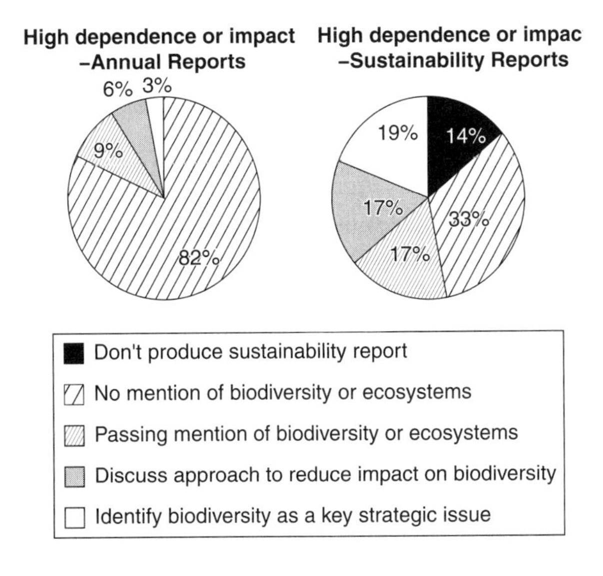

Figure 3.6 *Reporting by high impact or dependency sectors on biodiversity in 2008*

Source: PricewaterhouseCoopers for TEEB

issue (19 per cent versus 9 per cent) and a higher proportion reporting measures to reduce their impacts on biodiversity (36 per cent versus 24 per cent), compared with the top 100 companies.

More detailed examination of corporate reporting on BES shows that the information presented by companies is rarely sufficient to enable external stakeholders to form an accurate picture of companies' efforts to assess, avoid, mitigate or offset their impacts on BES. A survey conducted by Fauna and Flora International in 2008–2009, as part of the Natural Value Initiative, showed that companies in the food, beverage and tobacco sectors produced limited public disclosures, rarely stated clear targets and used mainly qualitative data (e.g. case studies, descriptions of initiatives) to communicate their management of biodiversity and ecosystems, rather than quantitative indicators of performance (Grigg et al. 2009). Only 15 of the 31 firms surveyed were able to provide reasonable disclosures in relation to BES, despite the focus of the survey on sectors in which both impacts and dependence on biodiversity and ecosystem services are considered relatively high.

Similar studies conducted by UK-based asset manager Insight Investment on the UK extractive industry and utilities, covering 22 companies in 2004 (Grigg et al. 2004) and 36 companies in 2005 (Foxall et al. 2005), produced similar results. Information on BES is often qualitative and frequently scattered throughout a company's website. As a result, it can be difficult for stakeholders, including investors, to assess whether a company understands its BES exposure and is managing these risks effectively.

These reviews suggest that even companies that are relatively advanced in their consideration of the issue are struggling to identify performance indicators for BES management and reporting. Several companies are working with multi-stakeholder initiatives, such as the Global Reporting Initiative sector supplements, to develop better standards of reporting (see Box 3.6). The development of sector-specific

Box 3.6 Selected initiatives offering guidance on BES measurement, management and reporting

Integrated Biodiversity Assessment Tool: www.ibatforbusiness.org
Natural Value Initiative: www.naturalvalueinitiative.org
Global Reporting Initiative – G3 guidelines and industry sector supplements: www.globalreporting.org
Stewardship Index for Specialty Crops: www.stewardshipindex.org
The Keystone Centre – Field to Markets Alliance for Sustainable Agriculture: www.keystone.org
Roundtable on Sustainable Palm Oil: www.rspo.org
Roundtable on Sustainable Biofuels: http://cgse.epfl.ch
Energy and Biodiversity Initiative: www.theebi.org
ICMM Good Practice Guidance for Mining and Biodiversity: www.icmm.com
IPIECA/API Oil and Gas Industry Guidance on Voluntary Sustainability Reporting: www.ipieca.org
WBCSD Cement Sustainability Initiative: www.wbcsdcement.org
Forest Footprint Disclosure Project: www.forestdisclosure.com
Water Footprint Network: www.waterfootprint.org

indicators offers the opportunity for more targeted measurement and reporting within sectors characterized by high BES impact or dependence.

3.5.3 Guidance on BES reporting

While few organizations in the public or private sectors report comprehensively (or at all) on biodiversity and/or ecosystem services in their annual report and accounts, a few more do so in separate annual sustainability or corporate responsibility reports. Here, unlike in financial reporting, there are as yet no mandated standards that all companies or organizations must follow. Examples of how some companies report on BES are provided in Boxes 3.7, 3.8, 3.9 and 3.10.

Box 3.7 Biodiversity reporting by Rio Tinto

Rio Tinto is a major international mining company, with operations in more than 50 countries, employing approximately 102,000 people. In 2004, Rio Tinto launched its biodiversity strategy, which includes the overarching goal to have a 'net positive impact' (NPI) on biodiversity. The company has developed practical tools and methodologies to assess the biodiversity values of their landholdings and has commenced, in association with its conservation partners, the application of offset methodologies in Madagascar, Australia and North America. In 2009, Rio Tinto completed a methodology for developing biodiversity action plans (BAPs) in collaboration with Fauna & Flora International (FFI). Rio Tinto reports on the biodiversity value of its sites, the amount of land in proximity to biodiversity-rich habitats, and the number of plant and animal species of conservation significance within these landholdings. This information is reported on their corporate website.

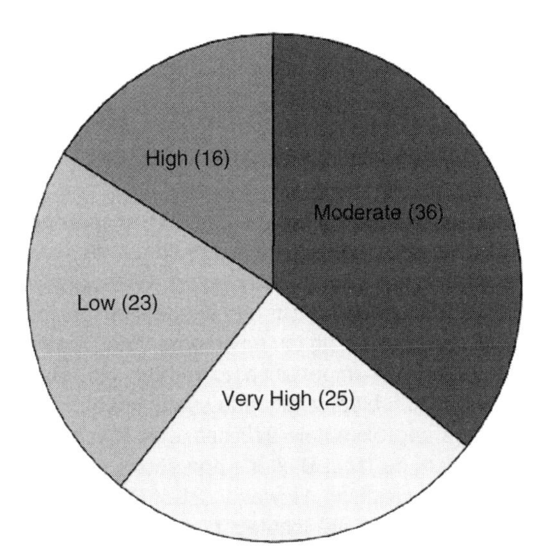

Source: Rio Tinto sustainability report (2009)

Box 3.8 Biodiversity in Scottish Power plc 2004 environmental report

Scottish Power publishes an annual Environmental Performance Report, which includes a section on 'Land and Biodiversity'. The 2004 report lays out the company's policy on biodiversity and summarizes their objectives and targets under five priority areas of minimizing impact, fisheries, birds, land reclamation and contamination.

Each section of Scottish Power's Environmental Performance Report sets out their potential impacts on biodiversity and their general approach to the issues. A key issue associated with birds, for example, is overhead power lines:

> *Overhead lines have the potential to harm birds so we have conducted surveys to identify high-risk areas, often in partnership with other organizations, and then implement bird protection programmes to reduce or prevent injuries and mortalities. In the UK bird diverters have been fitted to many lines crossing rivers and canals to reduce the risk of collision by wildfowl such as swans and geese.*

As well as summarizing the key biodiversity issues for the company and their achievements over the year, Scottish Power summarize their current targets, progress against these targets and set revised targets for the following year in concise tables. Progress is illustrated through boxes showing key achievements, such as awards received and new habitats created on Scottish Power land.

Source: Scottish Power environmental report (2004): www.scottishpower.com

Box 3.9 Biodiversity in Baxter Healthcare 2008 sustainability report

Baxter Healthcare has a long-standing commitment to transparency with respect to its environmental costs and savings. Baxter is also unusual in the level of detail provided on biodiversity issues in a recent sustainability report. While biodiversity is not among Baxter's stated *sustainability priorities,* it is an element of the company's bio-ethics policy: 'Baxter recognizes that protecting the environment and maintaining the biological diversity of our planet is of vital importance to human life. Baxter believes in the importance of maintaining global biodiversity and sustainable use of global resources.'

Baxter owns or leases approximately 910 hectares of land, about one-quarter of which is impermeable (paved) surface. Baxter's operations are typically located in light industrial areas in metropolitan regions. However, 21 of Baxter's 58 manufacturing and research and development facilities are located in some of the world's biodiversity 'hot spots', as defined by Conservation International.

Baxter facilities undertake various initiatives to protect biodiversity. For example, since 2006 the company's facility in Round Lake, Illinois, in the United States, has worked with a professional habitat restoration company and the local forest preserve to restore four hectares of the campus to a more natural habitat, including wooded savannah and a stream-bank riparian zone. As part of the project, the site used controlled burning to destroy invasive alien species. After the burn, Baxter employees planted 750 native plant species in an effort to facilitate regeneration of the native vegetation.

Source: Baxter sustainability report (2008): www.sustainability.baxter.com

Box 3.10 Environmental information in Eskom 2010 annual report

The South African power utility Eskom generates approximately 95 per cent of the electricity used in South Africa. The following extracts from the company's 2010 annual report focus on Eskom's environmental performance, with special reference to water, land and biodiversity. The report is compiled using the GRI guidelines and its Electric Utility Sector Supplement.

Governance

Environmental performance is managed as an integral part of our governance structure, from the board sustainability committee, to the executive management committee (Exco) sustainability and safety subcommittee. Accountable environmental managers and environmental practitioners from the various line divisions ensure the effective implementation of environmental management systems throughout our business. Corporate Services division sets overall Eskom strategy on the environment and provides oversight, reporting and assurance.

Impacts

Our activities that have significant environmental impacts include:

- *the construction of power stations and transmission and distribution power lines – impact on land use and ecosystems,*
- *the generation of electricity at our coal-fired power stations – use of resources (coal and water), land transformation, gaseous and particulate emissions and waste generation, and*
- *the generation of electricity at our nuclear power station – land transformation, radiation and radioactive waste.*

Responses

During this reporting year the Eskom land and biodiversity task team (represented by all the business areas within Eskom) has revised various biodiversity-related procedures, and drafted a biodiversity policy and standard within the framework of the ISO 14001 environmental management system. The Eskom–Endangered Wildlife Trust (EWT) strategic partnership performed a review of the effectiveness of the partnership, which highlighted the need to look at more pro-active and collective mechanisms to manage our impact on biodiversity. This led to the development of a partnership strategy with related KPAs and KPIs.

Performance (Generation Division)

- *Generation Peaking business unit achieved ISO 14001 certification*
- *Several other business units successfully completed phase 1 certification audits for ISO 14001*
- *Waste management reviews of all power stations were undertaken during the year to ensure improved waste management practices*
- *Water management and ground water reviews were completed to identify areas for improvement*
- *Water used as part of the process to generate electricity improved slightly from 1,35RA to 1,34RA L/kWh sent out.*

Performance (Transmission Division)

- *Development and implementation of biodiversity and land management environmental management plans for 330 existing power lines based on a phase-in approach: the proposed target of 90 per cent for 2010 was achieved (2009: 80 per cent)*
- *Transmission maintains nearly 29,000 km of overhead power lines, which means our divisional footprint needs to be monitored and managed closely through environmental management plans. The significant environmental impacts such as avian impact in terms of biodiversity are managed and controlled through our partnership with the EWT.*

Source: Eskom Holdings Ltd Integrated Report (2010): www.eskom.co.za

3.5.4 Integrated reporting

Increasingly, stakeholders are exploring how to integrate financial and non-financial information in a single report that provides a balanced and meaningful picture of a company. Early examples from companies such as Natura and Telefonica are based around providing annual reports and CSR/sustainability reports as a single package. Some companies produce these as paired documents and others as a single volume. In South Africa, the third King Code of Governance Principles, which took effect in 2010, emphasizes the importance of integrated reporting:

> Sustainability reporting and disclosure should be integrated with the company's financial reporting ... [t]he board should ensure that the positive and negative impacts of the company's operations and plans to improve the positives and eradicate or ameliorate the negatives in the financial year ahead are conveyed in the integrated report. (UNEP et al. 2010)

An international survey of corporate responsibility reporting in 2008, by KPMG, found that 86 per cent of the top 100 companies in South Africa by market value include some level of sustainability reporting in their annual reports. One company that includes such sustainability information in its annual report is the utility Eskom (see Box 3.10).

Alongside the pioneering efforts of individual companies, other networks and standards bodies are exploring how to promote more integrated reporting. In 2010, for example, an International Integrated Reporting Committee was launched by the Prince of Wales's Accounting for Sustainability Project (A4S) and the Global Reporting Initiative, with the aim to develop a 'globally accepted framework for accounting for sustainability ... a framework which brings together financial, environmental, social and governance information in a clear, concise, consistent and comparable format' (www.integratedreporting.org). It will do this by examining the interconnections between sustainability and financial factors in decisions that affect long-term performance, making clear the link with economic value.

Making the link between BES and financial indicators involves consideration of both input indicators and output indicators, where outputs may include damage to the environment. Traditionally, the latter aspect has been the only part of environmental performance that showed up in financial reporting, reflecting instances where companies were fined due to contravention of environmental regulations or liability for damages. More ambitious efforts to link BES and financial information would need to consider environmental inputs and outputs in terms of possible business risks and opportunities arising from changes in biodiversity and ecosystem services. Relevant information to support such reporting may include lists of species and their conservation status, protected areas and their extent, and other indicators of the environmental quality of water and soils. Ultimately, the reporting entity would adopt indicators that relate changes in environmental conditions to business operations, products and services. This implies, for example, the use of indicators that show the life cycle footprint associated with a company's products or the area of land under its management. The challenge is to consider also indirect impacts (see the earlier discussion on defining reporting boundaries and scope, and supply chain risks), and to assess the financial consequences of changes in environmental conditions.

The latter challenge should not be underestimated. It involves collecting, managing and tracking relevant information within a company and ensuring that the economic values of BES are properly reflected at a level of detail that can influence corporate financial analysis and decision making. This could be done, for example, by recording BES indicators for individual transactions (e.g. material purchase, sale, asset acquisition) in order to create accounting journal entries (Houdet et al. 2009). To help process this information, companies could make use of eXtensible Business Reporting Language (XBRL), a standard developed over the last decade to process and transmit financial data in a format that is computer-readable and easily usable for benchmarking purposes. This standard is ready for use, but its effectiveness for BES monitoring and evaluation depends on the degree of specificity that can be established for BES indicators, as well as the ability or willingness of those who are responsible for preparing company reports to use these indicators consistently.

3.5.5 Barriers to better BES accounting and reporting

Several barriers to comprehensive corporate disclosure on BES remain to be resolved, including:

- **Lack of consistent 'currency' or metrics:** although different initiatives and companies have developed metrics of relevance to BES, there is no single unit of measure or set of performance metrics that are consistently applied by companies within (let alone across) a range of sectors. In the case of greenhouse gas reporting by business, a key turning point was the publication of the WRI/WBCSD Greenhouse Gas Protocol (WRI/WBCSD 2004). Currently no such guidance exists for biodiversity or ecosystem service reporting. Developing such a standard is arguably more challenging than for greenhouse gases, as BES encompass a range of issues and there is no single indicator for assessing corporate processes or performance.
- **Perceived immateriality:** the absence of a compelling company-level business case that sets out the financial costs and benefits of (mis)managing BES results in a perceived low priority among many company managers and investors. The lack of clear regulatory guidance or market signals for many intangible ecosystem services is a large part of this problem.
- **Lack of understanding:** issues related to biodiversity and ecosystem services are often viewed as complex, in comparison with issues such as climate change or human rights. Understanding how BES issues and impacts relate to other sustainability concerns remains challenging and as a result many companies do not know how to measure and report effectively on BES.
- **Issues of scope:** the sustainable use of biodiversity and ecosystem services frequently goes beyond the boundaries of direct corporate ownership and control, and therefore is both difficult to measure and difficult to manage. Clarity is required on what constitutes a reasonable responsibility and reporting scope for a company.
- **Lack of demand:** the failure of investors to demand data on BES impacts and dependence may reflect the relatively short-term focus of the investment community.
- **Challenges in aggregation:** BES are often easier to measure for a specific site or area but difficult to aggregate into indicators for assessing overall organizational performance. Furthermore, it is often difficult to attribute changes in biodiversity or ecosystem services to the actions of an individual company.

3.6 Conclusions and recommendations

Despite all the barriers and challenges reviewed above, there are significant opportunities to improve the measurement and disclosure of BES in business. Companies can take action on their own or in collaboration with others to address these barriers. This section outlines the way forward for integrating BES in business valuation, accounting and reporting.

3.6.1 Technical improvements

Further work is needed on the science and practice of measuring BES at site, product and organizational levels. Opportunities include the following:

Advance scientific assessment, provision of relevant information and appropriate safeguards

Accessible information and relevant data on BES are essential for making sound business decisions that properly account for impacts on biodiversity and dependence on ecosystem services. Progress is being made in understanding how human actions affect the ecosystem, for example with satellite imaging and remote sensing. In addition, national and international standards are improving, and frameworks for assessing impacts on BES have been developed.

Gaps remain in scientific understanding of how the status of biodiversity or the condition of particular ecosystems affects the quality and quantity of ecosystem services, as well as how these may change if ecosystems are degraded. Some of the risks associated with ecosystem decline and biodiversity loss may be uncertain, but are potentially catastrophic and irreversible. An options valuation approach might be one way to address these risks – in effect, by preserving BES, a firm (and society) retains an option on the availability of these resources in the future. However, in view of the long-term considerations and present uncertainty about where tipping points lie, it may be difficult for firms to address these risks adequately or appropriately account for them. Final responsibility for avoiding catastrophic biodiversity loss may fall to government regulators, who are able to put limits on resource use or the conversion and disturbance of ecosystems. Moreover, governments, international organizations and other public bodies must work together to ensure that policy is communicated in a timely fashion, and that accompanying regulations are easy to understand and do not create perverse incentives.

Current impact assessment methodologies in life cycle assessment address a number of the drivers of biodiversity loss and ecosystem decline. Their results need to be communicated more widely so that LCA is able to provide an overview of a range of environmental impacts, including loss of BES. Advanced use of LCA faces barriers similar to those highlighted with respect to reporting methodologies, including understanding cause–effect chains, defining suitable indicators to quantify changes in BES and how to aggregate such indicators on a regional or global scale. Future research and methodological work on LCA and BES needs to expand the number of drivers of BES loss that are considered, develop consensus on impact categories and provide case studies that illustrate how impacts and associated damage can be translated into economic values. Research on land-use change can provide practical examples of the complexities involved and the type of methodologies able to link drivers of change in BES to loss through pressure indicators. Research on downstream risks and opportunities associated with BES, including product use and consumption, is still in its infancy.

In addition, life cycle management experts would do well to develop guidance for practitioners on BES boundaries and materiality, as they seek to add up impacts along the value chain. In particular, LCA experts need to provide guidance to help businesses identify the most material impacts and dependencies at different stages of the product life cycle and value chain. This requires, among other things, the use of inventory analysis to assess the relevance (direct/indirect) of the various components of biodiversity and different categories of ecosystem services within a particular industry. A tourism business, for example, has more direct links with the cultural services provided by ecosystems, whereas the food and beverage industries have clear dependence on provisioning services such as water.

Integrate BES information with core business planning and decision making systems

Methodological challenges relate not only to the choice of analytical framework (such as cost–benefit analysis), metrics (physical or financial) or the techniques used to value BES, but also to how information on BES values is to be integrated in business planning and decision making. It is important not to force 'mainstream' economic models into a business perspective, or superimpose a public economic approach onto business calculations. A more productive approach may be to find new ways of valuing BES impacts and dependencies within the context of existing financial and business planning procedures that companies already use. Unless BES values are considered by companies in the same way as other costs, benefits and management decisions, they are likely to remain marginal to corporate decision making.

3.6.2 Market improvements

There are many opportunities to improve valuation techniques and help markets recognize BES more effectively, four of which are summarized below.

Getting externalities into business valuation

When a firm's actions impose a significant externality, and existing legal avenues for redress are not sufficient, governments may wish to 'internalize' BES impacts into relevant business costs or revenues. In the case of damage to BES (a negative externality), taxes, licences or liability laws may be used to internalize the cost of adverse impacts. Similarly, tax credits or subsidies may be used to encourage business to conserve biodiversity or provide ecosystem services. Where appropriate, market-based instruments for BES (such as biodiversity credit and trading schemes) can be used to put a price on biodiversity impacts and dependencies (see Chapter 5 in this book).

Align business and social valuations, using regulatory and market mechanisms to reconcile differing discount rates

Assuming that most national economies will continue to have a large market sector, the challenge is to encourage commercial entities to make decisions that reflect the values of BES appropriately. Discounted cash flow analysis is likely to remain the dominant valuation and appraisal technique, and investors are likely to continue to expect managers to apply relatively high market discount rates in their financial analyses.

As with climate change, decisions taken today may have an impact on BES in the future as well as immediately. The challenge for regulators is to bring potential future losses of BES into today's business decision making. Policy makers will need to make judgements about the long-run costs of damage and define appropriate restrictions, licences or other incentives that will lead businesses to incorporate this cost in their investment decisions. If such policies are designed correctly, they can help to align business incentives with wider societal values.

Introduce techniques for capturing intangible values

When a business does recognize that its actions may have an impact on intangible BES values, there are several ways to measure this impact, such as contingent valuation. These tools are currently used most often by public policy makers to assess the social values of BES, but in future are likely also to be applied more widely by businesses, to evaluate their own impacts and dependencies.

More case studies are needed of the relevance of BES at the corporate level and for private investors (Box 3.11). Further research is required to establish the costs and benefits of managing BES sustainably (or mismanaging BES), with a view to establishing more reliable estimates that business (and regulators) can use to internalize biodiversity costs.

Box 3.11 The Ecosystem Services Benchmark

Several tools have been developed within the asset management community to evaluate the BES risks and opportunities of investments. One example is the Ecosystem Services Benchmark (ESB) tool, developed by the Natural Value Initiative in collaboration with investors from Europe, Brazil, the USA and Australia (Aviva Investors, F&C Investments, Insight Investment, Pax World, Grupo Santander Brazil and the Australian pension fund VicSuper). Designed to assess investment risks and opportunities associated with BES impacts and dependence in the food, beverage and tobacco sector, the Ecosystem Services Benchmark is aimed primarily at asset managers, but can also inform the banking and insurance sectors more generally. It has a secondary application for companies within the food, beverage and tobacco sectors, for which it provides a useful framework for planning and monitoring.

The ESB focuses on impacts and dependencies on biodiversity and ecosystem services associated with the production and harvesting of raw materials in companies with agricultural supply chains (including agricultural commodities, livestock and fish). It evaluates companies against broad categories of performance, such as competitive advantage, governance, policy and strategy, management, and implementation and reporting. Each company evaluated receives a summary of their results. By discussing the recommendations and outcomes of the analysis as part of routine investor dialogues with lower-ranked companies, improved performance can be encouraged and investor risk may be effectively managed.

Source: Natural Value Initiative (www.naturalvalueinitaitive.org)

Educate investors and set minimum requirements for BES in financial ratings

A review of 20 rating agencies, investment indices and ranking services conducted by IUCN (Mulder 2007) found only one that specifically referred to biodiversity – the Business in the Community Environment Index, in which companies were invited to answer voluntary questions on biodiversity. A more recent study by the Nyenrode Business School of Amsterdam showed that, while there is demand from clients such as pension funds, rating agencies rarely supply their clients with biodiversity-related information – in part because appropriate metrics are unavailable, but also because they see limited demand for such information. Ultimately, the quality of business measurement and reporting on BES will depend on the quality of questions asked by investors, analysts and other stakeholders.

3.6.3 Disclosure improvements

The information on BES presented in most company reports is rarely set out in a way that communicates that:

- key risks have been identified,
- policy and position on the issue is clear,
- a strategy to address those risks has been developed,
- management tools are in place to address the risks, and
- monitoring and review of processes is being undertaken to ensure implementation.

Without such information, reports on BES are of limited value to an investor or any other stakeholder with an interest in the issue. A recent benchmarking analysis of corporate disclosure of risks related to water scarcity, by CERES et al. (2010), concluded that the majority of companies in water-intensive industries exhibit weak management and disclosure of water-related risks and opportunities. The CERES report scored 100 companies based on five categories of disclosure: water accounting, risk assessment, direct operations, supply chain management and stakeholder engagement. Some steps that can be taken to improve disclosure of BES risks and opportunities are described below.

Encourage enhanced reporting on existing BES activities

Many companies could improve their reporting by disclosing more fully the actions they undertake to understand and manage BES. Examples include the presence of risk management frameworks, policies, strategy and targets, assessment of potential impacts on sensitive sites and actions taken to mitigate them, management plans and other activities undertaken to manage BES issues.

Increase cross-sector collaboration to develop and apply performance metrics and reporting guidance for BES

Pilot projects and cross-sector collaboration are required to develop better BES metrics that are relevant for both corporate management decisions and conservation monitoring. The need for sector-specific reporting guidelines within such an approach should continue to be explored. A number of platforms and processes exist at various levels that could advance the issue (e.g. the Convention on Biological Diversity, roundtables on sustainable soy, palm oil and other commodities, the GRI's ongoing work on reporting guidance for industry sectors supplements, etc.).

Improve mandatory requirements for companies to assess material environmental issues (including BES) in reporting

In many jurisdictions, companies are required to include material information in their annual accounts and are also sometimes subject to other forms of public reporting. With respect to annual accounts, governments should consider how to improve business understanding of materiality to ensure more comprehensive reporting on BES. Where economic instruments have been implemented to drive environmental performance (e.g. water quality or carbon trading), it is particularly important to provide such guidance.

In its overview of reporting legislation worldwide, the report *Carrots and Sticks for Starters* by UNEP and KPMG (2006: 57) noted that the effectiveness of regulatory instruments depends on the availability and quality of relevant information. For example, reliable reporting on greenhouse gas emissions is required for carbon markets to function properly. Many stakeholders would add that such disclosures require third-party verification and assurance, just as financial accounts require independent validation. The decision by the US Securities and Exchange Commission in early 2010 to require disclosure on climate change issues and the environment that affect capital expenditures (infrastructure issues), products and certain financial expenses (insurance) is likely to have significant bearing on future financial reporting of environmental assets and liabilities by business, not only in the USA.

In addition to reporting in annual accounts, other reporting requirements may be relevant, such as product-based, issue-based and site-based reporting. This raises the possibility of linking these requirements in a comprehensive reporting framework. The second *Carrots and Sticks* report, by UNEP et al. (2010), confirmed the growing regulatory interest in integrated corporate reporting. The trend towards greater integration of business reporting is likely to be accompanied by a proliferation of new tools for disclosure on greenhouse gas emissions, water use or other sustainability issues of interest to investors.

Acknowledgements

Wim Bartels (KPMG), Gerard Bos (Holcim), Sagarika Chatterjee (F&C Investments), Derek de la Harpe (African Conservation Projects Ltd), Frauke Fischer (Wurzburg University), Juan Gonzalez-Valero (Syngenta), Stefanie Hellweg (ETH Zurich), Kiyoshi Matsuda (Mitsubishi Chemicals), Narina Mnatsakanian (UNPRI), Herman Mulder (GRI), Kurt Ramin (IUCN), Virpi Stucki (Shell).

Notes

1 The report was originally published by ISIS Asset Management, a UK-based socially responsible investment firm that is now part of F&C Asset Management plc, under the title: 'Is Biodiversity a Material Risk for Companies? An assessment of the exposure of FTSE sectors to biodiversity risk' (September 2004). The report is available online at www.businessandbiodiversity.org/publications

2 More detail on corporate climate targets is provided under the UNEP Climate Neutral Initiative at: www.unep.org/climateneutral. Information on other companies' environmental commitments is available on their websites:
www.danone.com/en/press-releases/cp-octobre-2008.html
http://plana.marksandspencer.com/we-are-doing/climate-change
www.thecoca-colacompany.com/citizenship/water_main.html

www.riotinto.com/5273_biodiversity.asp
www.sony.net/SonyInfo/csr/eco/RoadToZero/

References

AccountAbility (2008) *AA1000 AccountAbility Principles Standard*, AccountAbility, London. URL: www.accountability21.net

Carson, R., Mitchell, R., Hanemann, W., Kopp, R., Presser, S. and Ruud, P. (1992) *A Contingent Valuation Study of Lost Passive Use Values Resulting from the Exxon Valdez Oil Spill*, Report to the Attorney General of the State of Alaska (November).

Carson, R., Wilks, L. and Imber, D. (1994) 'Valuing the preservation of Australia's Kakadu Conservation Zone,' *Oxford Economic Papers*, 46: 727–49.

CERES, UBS and Bloomberg (2010) *Murky Waters? Corporate Reporting on Water Risk*, CERES, Boston.

de Haes, U. et al. (1999) 'Best available practice regarding impact categories and category indicators in Life cycle impact assessment,' background document for the Second Working Group on Life Cycle Impact Assessment of SETAC-Europe (WIA-2), *International Journal of Life Cycle Assessment*, 4: 66–74.

Emerton, L., Grieg-Gran, M., Kallesoe, M. and MacGregor, J. (2005) 'Economics, the precautionary principle and natural resource management: key issues, tools and practices', in Rosie Cooney and Barney Dickson (eds) *Biodiversity and the Precautionary Principle: Risk and Uncertainty in Conservation and Sustainable Use*, Earthscan, London.

European Commission (2010) *International Reference Life Cycle Data System (ILCD) Handbook*, first edition, Publications Office of the European Union, Luxembourg (March).

Foxall, J., Grigg, A. and ten Kate, K. (2005) *Protecting Shareholder and Natural Value: 2005 Benchmark of Biodiversity Management Practices in the Extractive Industry*, Insight Investment, London.

GRI (2005) *GRI Boundary Protocol*, Global Reporting Initiative, Amsterdam.

GRI (2006) *Sustainability Reporting Guidelines*, Global Reporting Initiative, Amsterdam. URL: www.globalreporting.org

Grigg, A. and ten Kate, K. (2004) *Protecting Shareholder and Natural Value: Biodiversity Risk Management: Towards Best Practice for Extractive and Utility Companies*, Insight Investment, London. URL: www.naturalvalueinitiative.org/download/documents/Publications/PDF%203%20protecting_shareholder_and_natural_value2004.pdf

Grigg, A., Cullen, Z., Foxall, J. and Strumpf, R. (2009) *Linking Shareholder and Natural Value: Managing Biodiversity and Ecosystem Services Risk in Companies with an Agricultural Supply Chain*, Fauna and Flora International, UNEP FI and Fundação Getulio Vargas.

Hanson, C., Ranganathan, J., Iceland, C. and Finisdore, J. (2008) *The Corporate Ecosystem Services Review: Guidelines for Identifying Business Risks and Opportunities Arising from Ecosystem Change*, WRI, WBCSD and Meridian Institute, Washington, DC. URL: http://pdf.wri.org/corporate_ecosystem_services_review.pdf

Houdet, J., Pavageau, C., Trommetter, M. and Weber, J. (2009) 'Accounting for biodiversity and ecosystem services from a business perspective, Preliminary guidelines towards a Biodiversity Accountability Framework', Cahier no 2009-44, Ecole Polytechnique, Department of Economics.

Imber, D., Stevenson, G. and Wilks, L. (1991) *A Contingent Valuation Survey of the Kakadu Conservation Zone*, RAC Research Paper no. 3, Published for the Resource Assessment Commission, Australian Govt. Pub. Service, Canberra.

International Accounting Standards Board (IASB) (2001) *The Objective of Financial Statements,* IASB, London.

International Council on Mining and Metals (ICMM) (2003) 'Landmark 'no-go' pledge from leading mining companies'. URL: http://portal.unesco.org/culture/fr/files/12648/10614596949ICMM_Press_Relase.pdf/ICMM_Press_Relase.pdf

International Organization for Standardization (ISO) (2010) ISO 26000:2010, E. *Guidance for Social Responsibility*, ISO, Geneva.

ISIS Asset Management (2004) *Are Extractive Companies Compatible with Biodiversity? Extractive Industries and Biodiversity: A Survey*, ISIS Asset Management, London (February).

Jolliet, O., Margni, M., Charles, R., Humbert, S., Payet, J., Rebitzer, G. and Rosenbaum, R. (2003) 'IMPACT 2002+: A new life cycle impact assessment methodology', *International Journal of Life Cycle Assessment*, 8(6): 324–30 (November).

Jolliet, O., Miller-Wenk, R., Bare, J., Brent, A., Goedkoop, M., Heijungs, R., et al. (2004) 'The LCIA Midpoint-damage Framework of the UNEP/SETAC Life Cycle Initiative,' *International Journal of Life Cycle Assessment*, 9(6): 394–404.
JPMorgan Chase (n.d.) *Public Environmental Policy Statement*. URL: www.jpmorganchase.com/corporate/Corporate-Responsibility/environmental-policy.htm
King Committee (2009) *King III Report on Governance for South Africa*, Institute of Directors, Johannesburg.
Koellner, T. (2003) *Land Use in Product Life Cycles and Ecosystem Quality*, Peter Lang, Bern, Frankfurt a. M., New York.
Koellner, T. and Scholz, R. (2008) 'Assessment of land-use impacts on the natural environment, Part 2: Generic characterization factors for local species diversity in Central Europe,' *International Journal of Life Cycle Assessment*, 13(1): 32–48.
KPMG Sustainability B.V. (2008) *International Survey of Corporate Responsibility Reporting 2008*, KPMG International, Amsterdam.
Madsen, B., Carroll, N. and Moore Brands, K. (2010) *State of Biodiversity Markets Report: Offset and Compensation Programs Worldwide*. Ecosystem Marketplace, Washington, DC. URL: www.ecosystemmarketplace.com/documents/ acrobat/sbdmr.pdf
Mulder, I. (2007) *Biodiversity, the Next Challenge for Financial Institutions?* IUCN, Gland.
Oekom Research and Eurosif (2009) *Biodiversity: Theme Report – Second in a Series*, European Sustainable Investment Forum and Oekom Research AG, Paris and Munich.
Reid, Erin M. and Toffel, Michael W. (2009) 'Responding to Public and Private Politics: Corporate Disclosure of Climate Change Strategies,' *Strategic Management Journal*, 30: 1157–78 (DOI: 10.1002/smj.796).
Savage, D. and Jasch, C.M. (2005) *International Guidance Document on Environmental Management Accounting (EMA)*, International Federation of Accountants, IFAC, New York. URL: www.ifac.org
Schmidt, J. (2008) 'Development of LCIA characterization factors for land-use impacts on biodiversity,' *Journal of Cleaner Production*, 16(18): 1929–42.
TEEB Foundations (2010) *The Economics of Ecosystems and Biodiversity: Ecological and Economic Foundations* (ed. P. Kumar), Earthscan, London.
UK Environmental Agency (2004) *Corporate Environmental Governance: A Study into the Influence of Environmental Performance and Financial Performance*, Environment Agency, Bristol.
UNDSD (2001) *Environmental Management Accounting, Procedures and Principles*. United Nations Publications, New York, Geneva.
UNEP and KPMG (2006) *Carrots and Sticks for Starters: Current Trends and Approaches in Voluntary and Mandatory Standards for Sustainability Reporting*, KPMG, UNEP DTIE, Amsterdam and Paris.
UNEP, Global Reporting Initiative (GRI), KPMG Sustainability Services and University of Stellenbosch Business School (USB) (2010) *Carrots and Sticks: Promoting Transparency and Sustainability*, GRI, KPMG, UNEP, USB, Amsterdam.
Walker, S., Brower, A.L., Stephens, R.T. and Lee, W.G. (2009) 'Why bartering biodiversity fails,' *Conservation Letters*, 2(4): 149–57.
WBCSD (2011) *Guide to Corporate Ecosystem Valuation: A Framework for Improving Corporate Decision-Making*, WBCSD, Geneva. URL: www.wbcsd.org/web/cev.htm
WRI and WBCSD (2004) *The Greenhouse Gas Protocol*, WRI, WBCSD, Washington, DC, and Geneva. URL: www.ghgprotocol.org/standards/corporate-standard
Zadek, S. and Merme, M. (2003) *Redefining Materiality: Practice and Public Policy for Effective Corporate Reporting*, AccountAbility, London.

Chapter 4

Scaling Down Biodiversity and Ecosystem Risks to Business

Editors
Nicolas Bertrand (UNEP), Mikkel Kallesoe (WBCSD)

Contributing authors
Conrad Savy (CI), Bambi Semroc (CI), Gérard Bos (Holcim Ltd), Giulia Carbone (IUCN), Eduardo Escobedo (UNCTAD), Naoya Furuta (IUCN), Marcus Gilleard (Earthwatch Europe), Ivo Mulder (UNEP FI), Rashila Tong (Holcim Ltd)

Contents

Key messages

Biodiversity and ecosystem risks to business are real, tangible and should be managed. Public acceptance of biodiversity loss is declining, leading to calls for low-impact production and increased compensation for adverse impacts on biodiversity and ecosystem services.

Businesses are finding new ways to integrate biodiversity and ecosystem services (BES) into risk assessment and management. Many companies are exploring how to manage the adverse impacts of their activities on BES. A few companies have made public commitments to 'no net loss', 'ecological neutrality' or even 'net positive impact' on biodiversity, or on specific ecosystem services such as water supply.

A range of practical tools are available to help business reduce biodiversity and ecosystem risks. These include standards, frameworks and methodologies, data management tools, as well as model and scenario-building tools.

Economic valuation of biodiversity costs and benefits can inform risk management. In some cases, ecological restoration or offsets can deliver biodiversity and ecosystem benefits that exceed those of the original resource use, and at modest cost. More work is needed to integrate economic valuation into environmental risk management in business.

Managing biodiversity risk involves looking beyond sites and products to the wider land and seascape. In many industries, corporate environmental risk management has focused on direct or primary impacts. However, increasing public scrutiny and more stringent regulations have led companies in a range of sectors to extend their risk horizon to include indirect or secondary impacts.

Effective biodiversity and ecosystem risk management may be facilitated by appropriate enabling frameworks and partnerships. These may include new markets for biodiversity-friendly products, investment screening processes that require attention to biodiversity impacts and/or regulatory settings that pay close attention to biodiversity risks during the impact assessment process. Business risk management strategies also often involve public–private partnerships and stakeholder engagement.

4.1 Introduction: Biodiversity as business risk

Business risk can stem from uncertainty in financial markets, project failures, legal liabilities, credit risk, accidents, natural causes and disasters, as well as activist campaigns. Business risk management is a process of identifying potential risks, assessing the level of risk presented based on the likelihood and severity of potential impacts, and prioritizing responses based on available information and the relevance of the risk to the company.

Biodiversity loss and ecosystem degradation can also create business risks. Failure to manage these risks adequately can leave companies vulnerable to a range of potentially costly repercussions – including shortages of key inputs, advocacy campaigns, fines and increasing regulatory stringency, declining customer demand for products and services, or increasing difficulty in securing financing (see Chapter 2 in this volume).

Although current market practices do not adequately capture these BES risks, several factors suggest that this may be changing:

- Increasing requirements under voluntary and legislated codes of corporate governance as well as corporate social responsibility to disclose material risks and to develop systems for managing such risks.[1] Sometimes these include risks related to biodiversity and ecosystem services.
- New environmental legislation, such as the EU Environmental Liability Directive (ELD), which increases potential liabilities of business and therefore may create the need for new risk transfer solutions, such as insurance. As a consequence, insurers are exploring new ways of evaluating BES risks and potential new insurance products (Busenhart et al. 2010).
- The integration of sustainability issues, including, in some cases, BES in business school curricula and other business training initiatives.[2]

Many BES tools have been developed that can help companies identify potential risks, assess the degree of risk and develop effective management strategies to mitigate them. However, the integration of BES within corporate risk management systems is still in its infancy, partly due to limited scientific understanding of BES, which can make it difficult to predict the likelihood or severity of business impacts. Nevertheless, some far-sighted companies have discovered that BES risks can also be seen as an opportunity to increase market share or to develop new or improved products and services. In short, good management can turn a BES risk into an asset rather than a liability (for more details see Chapter 5 in this volume).

This chapter explores actions taken by companies to identify, manage and reduce their biodiversity and ecosystem risks. It begins with an overview of corporate risk management systems and efforts made to integrate BES within them (Section 4.2), followed by a review of existing and emerging tools that can help business identify and manage these risks (4.3). The chapter continues with a discussion of best practices for risk management, including stakeholder engagement and adaptive management (4.4), followed by conclusions and recommendations for action (4.5).

4.2 Integrating biodiversity and ecosystem services into corporate risk management

A standard risk management framework characterizes risk according to the degree of potential impact and the likelihood of the impact occurring. In its simplest form, a two-dimensional matrix of probability (likelihood) and consequence (severity) can be used to evaluate risk. More often, however, the complexity of the issues involved calls for more elaborate representation, with 4 × 4, 4 × 5 or 5 × 5 risk matrices used.

The cells of such a matrix are typically assigned to one of several risk classes (e.g. critical, very high, high, moderate and low) and may be interpreted as priorities to which specific management responses are attached. A low probability/high impact risk may thus be put in the same risk category as a moderate probability/moderate impact risk. In addition to determining an acceptable level of risk, a company will often develop a risk management strategy that identifies residual risks, develops a control strategy and assigns accountability to manage these (Jones and Sutherland 1999).

Building on the examples provided in Chapter 2, we illustrate below how biodiversity loss and ecosystem degradation can create risks to companies:

- The electric power sector faces a range of business risks linked to climate change and ecosystem degradation. A total of 74GW of generating capacity, or over half of existing and planned capacity for major power companies in Asia, is located in areas that are considered water-scarce or water-stressed. Overuse of water or degradation of watersheds, such that they are less able to regulate water flows, can disrupt the operations of power generation facilities, causing temporary interruptions in electric power or even permanent losses of capacity, as well as lower revenues and higher costs (Sauer et al. 2010).
- The petroleum industry is increasingly forced to explore and produce in ever more sensitive environments, as traditional oil-producing regions mature and yield progressively less oil. In some cases, access to petroleum reserves may be denied, restricted or contested. Even when access is legally allowed, opposition from local communities can seriously hamper production. Austin and Sauer (2002) assessed the financial implications of possible restrictions on access by extractive companies to proven reserves in ecologically important and protected areas. Under different scenarios, access pressures were reflected by different combinations of increased production costs, reduced production capacity and reserves being placed 'off-limits', with knock-on effects on shareholder value (Figure 4.1).

More generally, a summary of BES-related risks for various sectors is presented in Table 4.1.

Integrating biodiversity and ecosystem service risks into the standard corporate risk management framework presents some unique challenges. One issue, discussed at length in Chapter 3, is that companies generally base their risk assessments on the potential harm to their business interests and account for them in financial terms. Risks that do not have a direct financial impact on the company or that are more difficult to translate into economic losses and gains, such as impacts on biodiversity and ecosystems, are seldom quantified or monetized (Houdet 2008). Such risks may

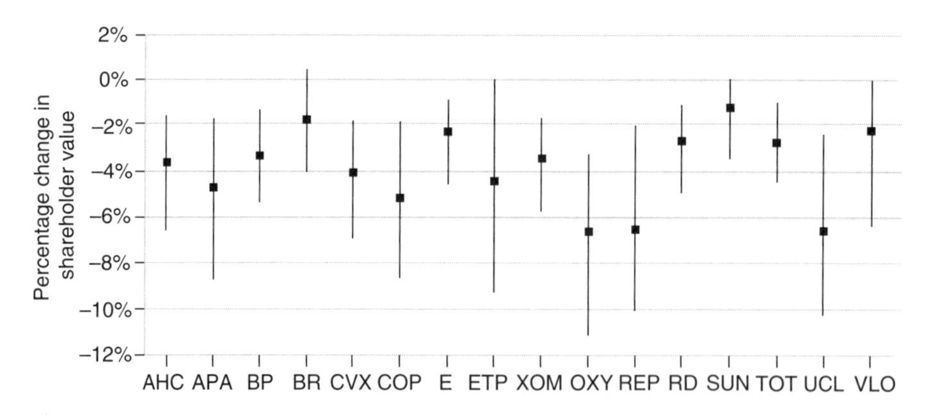

Figure 4.1 *Financial consequences of climate policies and restricted access to petroleum reserves*

Source: Austin and Sauer (2002)

Note: The graph shows the range of possible outcomes and most likely impact for 16 companies: Amerada Hess (AHC), Apache (APA), BP (BP), Burlington Resources (BR), ChevronTexaco (CVX), Conoco- Phillips (COP), Eni (E), Enterprise Oil (ETP), ExxonMobil (XOM), Occidental Petroleum (OXY), Repsol YPF (REP), Royal Dutch/Shell Group (RD), Sunoco (SUN), TotalFinaElf (TOT), Unocal (UCL), and Valero Energy (VLO).

Table 4.1 Biodiversity risks in different sectors

		Sectors most likely to be affected							
		Primary industries (e.g. forestry, oil and gas, mining, farming and fishing)	**Utilities** (e.g. electricity, gas, water)	**Consumer goods** (e.g. automobiles, food products, household products)	**Consumer services** (e.g. retailers, media, travel and leisure)	**Health care** (e.g. pharmaceuticals, biotechnology, health care providers)	**Industrials** (e.g. construction, aerospace, components)	**Financials** (e.g. banking, insurance, asset management)	**Technology and business services** (e.g. software, telecoms, consulting)
Category	**Risks**								
Operational The day-to-day activities, expenditures and processes of the company	Increased scarcity or costs of inputs; reduced quality of inputs	•	•	•	•	•	•		
	Reduced output or productivity	•	•	•					
	Disruption to business operations	•	•	•	•	•	•	•	•
	Supply chain risks			•	•	•			

Table 4.1 Biodiversity risks in different sectors *(Cont'd)*

		Sectors most likely to be affected							
		Primary industries (e.g. forestry, oil and gas, mining, farming and fishing)	**Utilities** (e.g. electricity, gas, water)	**Consumer goods** (e.g. automobiles, food products, household products)	**Consumer services** (e.g. retailers, media, travel and leisure)	**Health care** (e.g. pharmaceuticals, biotechnology, health care providers)	**Industrials** (e.g. construction, aerospace, components)	**Financials** (e.g. banking, insurance, asset management)	**Technology and business services** (e.g. software, telecoms, consulting)
Regulatory and legal The laws, government policies and court actions that can affect corporate performance	Restricted access to land and resources (e.g. extraction moratoria, permit or licence suspension, permit denial)	•	•						
	Litigation (e.g. fines, lawsuits)	•				•			
	Lower quotas	•							
	Pricing and compensation regimes (e.g. user fees)	•	•				•		
Reputational The company's brand, image or relationship with customers, the general public and other stakeholders	Damage to brand or image; challenge to 'licence to operate'	•	•	•	•	•	•	•	•
Market and products Product and service offerings, customer preferences, and other market factors than can affect corporate performance	Changes in consumer preferences	•		•	•				
	Purchaser requirements	•		•					
Financing Cost and availability of capital from investors	Higher cost of capital; more rigorous lending requirements	•	•					•	

Source: Adapted from Evison and Knight (2010) and Hanson et al. (2008)

be perceived as less tangible or having only an indirect impact on the health of a company. In some instances, however, they may result in a very direct impact on a company's licence to operate and, in the worst cases, lead to shutdowns and direct economic losses. The challenge is to bring BES and other non-market risks into business accounting frameworks and decision making processes, so they are considered alongside traditional risk factors.

Although some studies have attempted to link reputation, consumer confidence and trust to brand value, it is not currently possible to determine the impact of effective biodiversity risk management on the value of companies or brands (Earthwatch 2002; F&C Investments 2004). Instead, many companies simply refer to 'doing the right thing' to account for their voluntary investments in biodiversity conservation (i.e. efforts that go beyond statutory requirements).

A related challenge is that public policy and expectations about biodiversity and ecosystem services are rarely expressed in economic terms. Typically, the values of biodiversity and ecosystem services are implicit rather than explicit, with risk defined in terms of government regulations (e.g. Natura 2000 sites), scientific consensus (e.g. the IUCN Red List of Endangered Species) or consumer preferences (e.g. certified 'sustainable' products). While such expressions may include clear limits or thresholds of acceptable risk, these are often defined in terms of inconsistent bio-physical units, such as species populations or ecosystem extent. Integrating such metrics into business risk management systems can be extremely challenging, especially when there are conflicting views on the relative importance of, or risks to, particular species, sites or ecosystems.

A third challenge is how to deal with indirect risks. Most companies, when assessing risks, tend to focus on direct or primary impacts ('within the fence') which can be avoided or mitigated through changes in operating procedures (EBI 2003). However, growing public concern, stricter regulation and more stringent voluntary standards have led an increasing number of companies to extend the boundaries of risk management beyond their direct impacts to include indirect or secondary impacts.

Indirect impacts do not result from business projects or operations themselves but may be triggered by the presence of a project or company in the wider landscape (EBI 2003). For example, the presence of a mine in a remote area may lead to immigration, which in turn can result in land-use change or increased water use far beyond the area directly affected by mining operations. Growing awareness of economic and ecological linkages and the potential for cascading biodiversity and ecosystem effects across a wide area has led to increasing interest in landscape-level assessments, life-cycle assessment processes and supply chain management strategies designed to identify and mitigate these risks. Such concerns are reflected, for example, in recent guidance on integrating biodiversity in management systems in the oil and gas and mining sectors (EBI 2003; IPIECA and OGP 2005; Johnson 2006). Similarly, a growing number of product certification standards developed within the agriculture, fisheries, aquaculture and forestry sectors now include landscape-level biodiversity indicators.

In many cases, biodiversity and ecosystem service risks are influenced by external factors that are not under a company's total control, making them difficult to predict and to manage. Reputational risks, for instance, are created by the perception of external stakeholders on how business is conducted. Trust and confidence in a company, or lack thereof, is often driven by public perceptions of the direct or indirect impacts of a company's activities, which are not always well informed. Similarly, regulatory restrictions may result from societal pressure to prevent real or imagined impacts, as in the case of genetically modified organisms. While there is value for business in anticipating reputational risks and taking advantage of opportunities to inform new regulations, the benefits of doing so can be difficult to quantify.

Because BES risks are often indirect and emerge from factors beyond the control of a particular company, an effective strategy for managing these risks will involve a shared responsibility among governments, NGOs and business decision makers. In addition, effective management of BES risks in business will probably entail the use of an expanded framework that includes both direct and indirect impacts on biodiversity and relates these (to the extent possible) to the financial health of the company. This will often require detailed analysis and broad stakeholder engagement to enable business to assess the full value of their impacts on biodiversity (see Section 4.4.1).

An example of how business can work with partners to assess biodiversity risk is provided by the cement and construction company Holcim, which has developed a biodiversity risk matrix in collaboration with IUCN and a panel of experts (see Box 4.1).

Box 4.1 Holcim and IUCN: Implementing a biodiversity management system

In 2007, Holcim began working with IUCN in order to understand better the risks and opportunities facing the company in relation to biodiversity. The relationship has helped Holcim develop a corporate approach to the issue and to define biodiversity-related activities at site level over the full life cycle of its operations. One result is a new biodiversity management system (BMS), which is used to assess biodiversity issues in new projects and to determine appropriate corrective actions in sites of varying environmental sensitivity.

A key step in the BMS is the establishment of a biodiversity risk matrix, followed by the introduction of measures appropriate for the level of risk encountered at each site (see figure). The risk level is determined first in terms of biodiversity value or importance (i.e. proximity to high biodiversity value areas) and second by the potential direct impact of Holcim operations. This methodology also takes into account the biodiversity value expressed by relevant local stakeholders. On this basis, certain sites are categorized as 'sensitive': namely sites of national or global importance for biodiversity conservation where operational impacts are considered to be 'very high', 'high' or 'medium'.

Biodiversity Importance	Potential Impact			
	Very High	High	Medium	Low
Global	Critical	Significant	Medium	Low
National	Critical	Significant	Medium	Low
Local	Medium	Medium	Low	Low
Low	Low	Low	Low	Low

Source: IUCN–Holcim independent expert panel (adapted by Holcim; © Holcim 2010)

The matrix is used as part of three implementation steps in the BMS:

- *Step 1. Know the potential impact:* an annual environmental questionnaire is used to collect (self-reported) biodiversity information at site level and used for risk mapping. Where risks or impacts are unknown, the knowledge gap is flagged.
- *Step 2. Match the level of effort to risk*: sites categorized as sensitive (see above) are required to implement biodiversity action plans and monitor progress. Expert partners may be enlisted to help conduct biodiversity inventories as needed, set appropriate targets and determine actions.
- *Step 3. Monitor results to demonstrate progress towards the targets:* at most Holcim sites, monitoring can be conducted by company staff. For sensitive sites, however, external monitoring can provide additional credibility. Biodiversity activities need to be integrated into existing operational management processes, such as rehabilitation planning and environmental management systems.

An initial inventory of all 500+ extraction sites owned by Holcim in over 70 countries was conducted and all sites were categorized using the risk matrix. Senior managers were informed about which sites needed attention first and a global target was set in order to monitor progress: by 2013, 80 per cent of sensitive sites will have a biodiversity action plan in place. Progress will be published in the company's sustainability reports and Holcim will continue to work with external partners, where appropriate, while also building capacity internally to assess biodiversity and monitor progress.

Sources: Carbone et al. (2009b); Holcim – Bos and Tong (2010)

Extensions of the standard corporate risk management framework to address biodiversity and ecosystem services have typically built on established systems and processes, including risk registers, environmental impact assessment (EIA) and environmental management systems. These often emphasize certain high impact stages within the project life cycle (Figure 4.2), notably early stages when costs may be lower and responses are most flexible. Many companies, however, attempt to manage risk throughout the project life cycle, including the operational phase. Some companies find that their greatest BES risks are associated with the sourcing of raw materials, and therefore adopt or develop improved supply chain management systems to mitigate these risks as well.

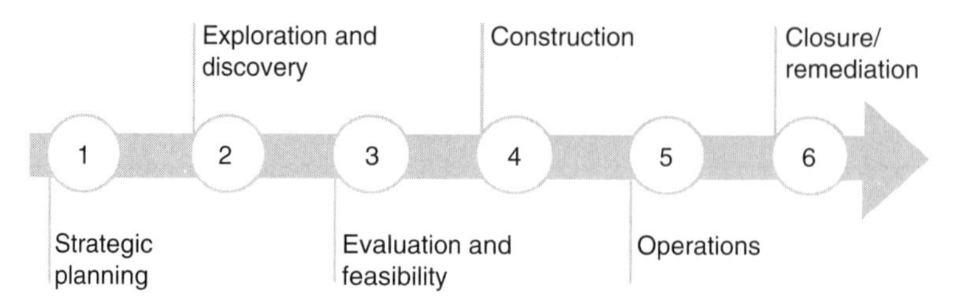

Figure 4.2 *The extractive project life cycle*

Source: Conservation International (2010)

Where biodiversity and ecosystem service impacts are not well reflected in risk assessment and management (e.g. due to weak legislation, lack of public awareness or limited scientific knowledge), the use of economic valuation can help clarify the values at stake and raise awareness of the magnitude of potential BES risks. In this regard, a challenge for business is that the methods used to estimate the economic values of biodiversity and ecosystem services have not been adapted to business needs in the same way; for example, that biodiversity surveys and assessment methods have been adapted to and incorporated in corporate environmental impact assessment or product environmental standards.

Existing corporate risk management frameworks tend to manage risks using a simple hierarchy that is designed to avoid the highest-risk impacts to the extent possible and to minimize and mitigate those risks that cannot be avoided (BBOP 2009a). Although the environmental mitigation hierarchy typically takes little account of economic benefits, it is nevertheless well suited to managing BES risks. It enables companies to consider options for managing risks associated with potential impacts and whether these can be avoided or should be part of a mitigation plan designed to reduce or compensate for adverse impacts. Companies are increasingly exploring ways to manage the unavoidable, residual impacts that are inherent in this approach, through the use of biodiversity offsets (Box 4.2).

Box 4.2 Applying the mitigation hierarchy and biodiversity offsets in the financial sector

The United Nations Environment Programme Finance Initiative (UNEP FI) and the Business and Biodiversity Offsets Programme (BBOP) recently commissioned the sustainability and climate change team at PricewaterhouseCoopers to conduct a study exploring the following themes:

- familiarity and awareness in the financial sector of the biodiversity mitigation hierarchy and biodiversity offsets;
- corporate policy approaches of banks to understanding and addressing biodiversity issues;
- roles and responsibilities within the banks on the management of biodiversity risks and opportunities; and
- tools, resources and training.

Based on 26 in-depth discussions with banks, environmental consultants, NGOs and bank clients, the analysis concluded that:

- Biodiversity issues do affect financing decisions.
- Business drivers for banks are currently risk based and primarily reputational.
- While the implementation of a biodiversity management framework is largely focused at the asset level, the application of the mitigation hierarchy and offsets is in its infancy.
- Bank policy frameworks and procedures are evolving to incorporate a broader, strategic consideration of biodiversity risks.

Source: PricewaterhouseCoopers (2010)

The mining company Rio Tinto, for example, set out a biodiversity strategy in 2004 that includes a long-term goal of net positive impact on biodiversity. This means ensuring, where possible, that the company's actions have positive effects on biodiversity features and values that not only balance but are broadly accepted to outweigh the inevitable negative effects of the physical disturbances and impacts associated with mining and mineral processing (Rio Tinto 2008). Rio Tinto aims to achieve this by reducing impacts and implementing positive conservation actions in the form of biodiversity offsets and other conservation measures (Figure 4.3).

The concept of biodiversity offsets involves tying the level of compensation or investment in biodiversity conservation to the level of impact of a project or business. There is a clear analogy with carbon offsets, which are typically linked to the level of greenhouse gas emissions of an individual or entity. The Business and Biodiversity Offsets Programme (BBOP) has defined biodiversity offsets as:

> Measurable conservation outcomes resulting from actions designed to compensate for significant residual adverse biodiversity impacts arising from project development after appropriate prevention and mitigation measures have been taken. The goal of biodiversity offsets is to achieve no net loss and preferably a net gain of biodiversity on the ground with respect to species composition, habitat structure, ecosystem function and people's use and cultural values associated with biodiversity. (BBOP 2009b)

Based on the definition put forward by BBOP, companies should not only avoid and mitigate their adverse impacts on biodiversity, but should also take steps to measure the residual impacts of their operations on biodiversity and ensure that these are offset

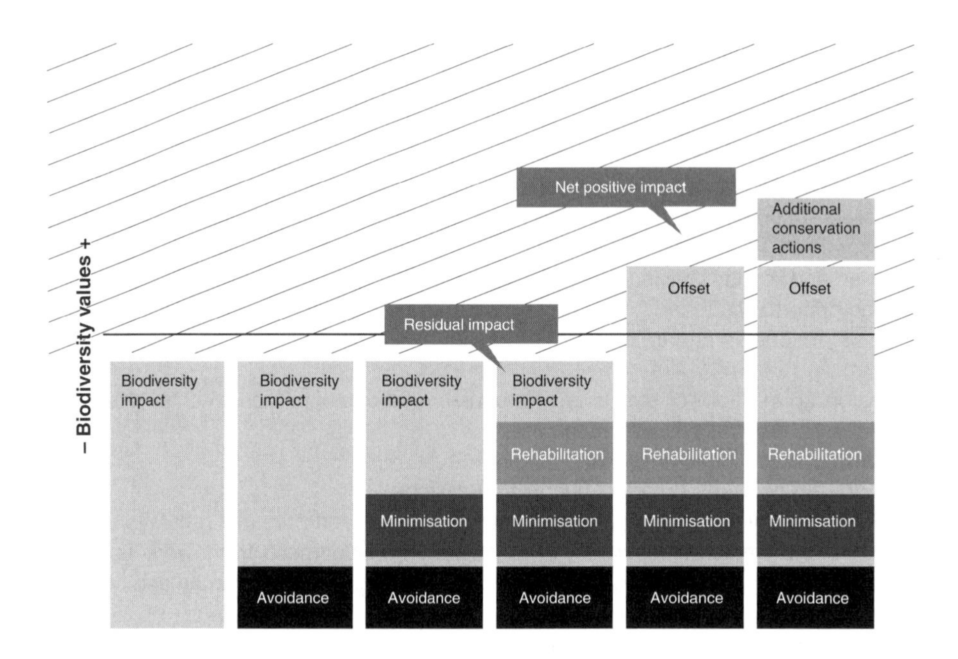

Figure 4.3 *Net positive impact and the mitigation hierarchy*

Source: Rio Tinto (2008)

through conservation investments in comparable sites. In practice, assessing biodiversity and ecosystem impacts has been hampered by the costs of collecting data and incomplete knowledge of BES. It can be difficult to define an acceptable impact in terms of ecosystem and species resilience, or to anticipate and avoid all irreversible impacts. In addition, most biodiversity offsets to date have been designed using purely bio-physical indicators of biodiversity value, although BBOP acknowledges the importance of local use values and provides guidance on how to measure them. If biodiversity offsets do become standard practice, and if economic valuation is used as part of the assessment of biodiversity impacts and offsets, the approach could help integrate the economic values of biodiversity and ecosystem services in industrial development and risk management processes. More generally, economic valuation can also be applied in the context of the mitigation hierarchy, to assess the value of biodiversity impacts and the relative costs (or cost-effectiveness) of avoidance, mitigation and offsets.

4.3 Tools for managing biodiversity and ecosystem risks

A variety of tools have been developed to help businesses identify and manage their biodiversity and ecosystem service risks (see OECD 2005 and Waage et al. 2008 for two recent surveys). Typologies based on which tools fit where in the different stages of risk assessment or the project life cycle have certainly facilitated uptake. However, a broader perspective on risk management tools can also be useful:

- Some tools, such as ISO 14001 or the Corporate Ecosystem Services Review (ESR), take a broad strategic look at risk assessment and can be applied at various stages in a project life cycle.
- The project life cycle itself is usually an iterative process, involving adaptive management, testing and feedbacks.
- Reliance on a single 'tool' at a specific step in the risk assessment process may not be sufficient to support the full set of business needs at that specific stage.

The typology presented below takes a different approach and examines three broad types of tools: standards, frameworks and methods; data management tools; and model- or scenario-based tools.

4.3.1 Standards, frameworks and methodologies

Standards, frameworks and methodologies include a wide range of tools for BES risk management. Standards establish 'rules, guidelines or characteristics for products or related processes and production methods' (ISEAL 2010). Standards set the bar for business (or their customers) to determine whether they have achieved a certain level of 'best practice', in terms of good processes (such as environmental management) or good performance (such as avoiding critical habitats). Examples of voluntary, sector-based or cross-sector standards for business include ISO 14001 for environmental management (ISO 2004), the Programme for the Endorsement of Forest Certification (www.pefc.org), the Forest Stewardship Council (www.fsc.org), and the WBCSD Greenhouse Gas Protocol (www.ghgprotocol.org). Individual company standards include Starbucks' Coffee and Farmer Equity (CAFE) practices, which set criteria and indicators for environmentally and socially responsible coffee purchasing and are used to verify performance throughout the company's supply chain (www.starbucks.com/responsibility/sourcing/coffee).

Depending on the nature and the size of the company, internal standards can have significant impacts on wider supply chains, and even entire sectors. For example:

- Walmart's sustainability initiatives have arguably influenced previously intractable value chains, such as for gold and gemstones (www.loveearthinfo.com);
- The International Finance Corporation (IFC) Performance Standard 6 (PS6) on Biodiversity Conservation and Sustainable Natural Resource Management identifies when projects or investments may occur in high risk areas and outlines a series of steps to minimize these risks. PS6 has had significant impact on project financing more generally, by virtue of being adopted as part of the Equator Bank Principles and applied by some 70 major financial institutions in all projects in excess of US$10 million (www.equator-principles.com).

Internal company standards are often developed to ensure that managers are aware of risks and that these are measured consistently. Especially where government regulations are absent, weak or unenforced, conventional indicators of poor BES performance (e.g. prosecutions, licence violations) will not be sufficient to inform company managers and owners of how the company is managing risk. Corporate standards backed by external auditing and reporting provide additional assurance that potential risks are being identified and addressed.

Key components of these systems are the establishment of clear principles and criteria that present the level of performance expected of a supplier or operation, indicators to monitor performance over time, and a verification process (ideally third-party) to ensure the validity of claims.

The process of transforming good practice standards into formal certification systems is often lengthy and costly, and may not be warranted in all cases. For instance, many businesses have developed internal guidance on biodiversity and ecosystem services and implement these as part of their standard environmental management systems, using internal verification only. In other cases, additional tools may be needed to fulfil the requirements of a standard or guideline. In such cases, supplemental methods and frameworks might be developed to define a suggested set of activities, provide checklists or other recommendations, to help a business assess its performance against a particular BES standard or objective:

- The IFC, for example, has developed notes to support implementation of their standards. Similarly, BBOP is developing additional guidance to help practitioners interpret their suggested principles, criteria and indicators for 'good' biodiversity offsets, based on lessons learned from pilot projects.
- Supplemental guidance may be explicitly tied to an existing standard, such as the High Conservation Value (HCV) framework, which was originally developed to support Forest Stewardship Council (FSC) certification. This framework defines biodiversity and ecosystems according to six broad areas of concern, including ecological, social, cultural and livelihood values.
- In other cases, guidance may stand independently of other standards and support a general intent to understand biodiversity and ecosystem service risks and

values. This is the case for the Corporate Ecosystem Services Review (ESR), developed by the World Resource Institute (WRI), the World Business Council for Sustainable Development (WBCSD) and the Meridian Institute (Hanson et al. 2008). The ESR provides step-by-step guidance on how businesses can identify and manage risks related to ecosystem services, including those they impact as well as those they depend on for their operations (see Box 4.3). In such cases, guidance may support multiple standards or policies or could form the basis of entirely new standards.

Box 4.3 The Corporate Ecosystem Services Review

The Corporate Ecosystem Services Review (ESR) is a structured methodology for managers to develop strategies for managing business risks and opportunities arising from their company's dependence and impacts on ecosystems. It helps companies to identify impacts and dependence on healthy ecosystems such as fresh water, timber, genetic resources, pollination, climate regulation and natural hazard protection, and connects these to their bottom line. The ESR builds on existing environmental management systems and due diligence tools by:

- looking beyond issues of pollution and natural resource consumption;
- addressing dependencies and impacts companies have on the natural environment;
- considering both business opportunities and risks;
- coupling economic and environmental issues;
- providing a framework for stakeholder engagement; and
- enabling use of environmental risks and opportunities for more innovative corporate strategy.

The ESR has been applied by an estimated 200 companies and has delivered many benefits through the following means.

1 Enabling companies to identify new business opportunities arising from their dependence and impact on ecosystems and the services they provide, and to anticipate new markets as they are developing. In this way, the tool has also provided companies with information necessary to influence government policies on ecosystem conservation.
2 Strengthening existing approaches to environmental impact assessment by addressing ecosystem issues not usually considered during that process. For instance, in the case of Mondi, Europe's largest producer of office paper, application of the ESR led to the development of initiatives to improve water efficiency through control of invasive species and selection of water-efficient tree strains.
3 Providing a framework in which stakeholder engagement processes and relationships can be improved. Syngenta, a large agribusiness company, applied the ESR to identify opportunities to engage a growing customer segment in India and identified multiple opportunities to provide additional services to these farmers.
4 Enabling companies to demonstrate leadership in this area by addressing the decline in ecosystem services.

Source: Hanson et al. (2008)

4.3.2 Data management tools

The existence of standards, frameworks and methodologies, as outlined above, goes a long way to help businesses identify and manage their BES risks. Access to data and reliable interpretations of statistical information are essential to make such tools useful.

In most cases, businesses have relied on context-specific data drawn from original studies that are guided by established frameworks or methods for data collection and analysis (e.g. environmental impact assessment). However, business sustainability frameworks and standards are increasingly designed to draw on existing data collection processes and tools. This helps reduce costs, but also makes it easier to draw on existing expertise in the scientific or NGO community. The Roundtable on Sustainable Palm Oil (RSPO), for instance, makes reference to the High Conservation Value (HCV) framework in its description of sensitive habitats. HCV assessments are informed by national toolkits, where available, and have traditionally drawn on national data to map these areas.

Access to supporting data for such efforts can be challenging, due to the abundant, dispersed and often difficult-to-access nature of national BES data. The Integrated Biodiversity Assessment Tool (IBAT) was developed as one solution to such needs, with the aim of offering rapid, easy access to the most critical fine-scale national datasets which indicate critical habitats, protected and unprotected areas (Box 4.4).

Box 4.4 Integrated Biodiversity Assessment Tool

The Integrated Biodiversity Assessment Tool (IBAT) for business is a partnership between BirdLife International, Conservation International, IUCN and UNEP–WCMC, which aims to facilitate corporate access to information on globally recognized, site-level biodiversity priorities, in order to assess potential risks at the earliest stages of decision making within the project life cycle. IBAT is intended to provide the most recent available data on sites of global conservation importance (i.e. protected areas and Key Biodiversity Areas) on a web-based platform, thereby allowing companies to conserve biodiversity and minimize corporate risks by developing plans to avoid or mitigate potential adverse biodiversity impacts.

Source: www.ibatforbusiness.org

Another example of a data management tool for business is the Global Water Tool (GWT), which uses pre-loaded datasets to calculate water-related indicators for the Global Reporting Initiative (Box 4.5).

Box 4.5 The Global Water Tool: Helping companies make water-informed decisions

The Global Water Tool (GWT), developed by WBCSD member company CH2M HILL in partnership with 22 other companies, the Nature Conservancy and GRI, helps companies and organizations map their water use and assess corporate water risk across their global operations and supply chains. The GWT can be used to:

- Compare a company's water uses (including staff presence, industrial use and supply chain) with key external water-related data.
- Create key water Global Reporting Initiative (GRI) indicators, inventories, risk and performance metrics and geographic mapping. The GRI indicators on total water withdrawals (EN8), water recycled/reused (EN10) and total water discharge (EN21) are calculated for each site, country, region and in total.
- Establish relative water risks in a company's portfolio to prioritize action.
- Create graphs and maps.
- View facilities spatially through Google Earth, which provides detailed geographic information, including surface water.
- Enable effective communication with internal and external stakeholders on a company's water issues.
- Calculate water use and efficiency.

A significant limitation of the GWT is the quality of the underlying water datasets. For this reason, the tool does not aim to provide specific guidance on local situations, which usually require more in-depth analysis. Other tools, such as the Global Environmental Management Initiative Water Sustainability Planner Tool, may be more targeted towards this outcome.

About 300 companies have used the GWT to date. Some companies use it as a first step to identify water 'hot spots', where more in-depth analysis is needed. The Dow Chemical Company, for example, used the GWT to map its 157 manufacturing sites around the world, drawing attention to water supply risks at several sites. Similarly, Borealis used the GWT to make water projections across a range of sites to 2025, as part of long-term planning for sustainable water management.

Source: www.wbcsd.org/web/watertool.htm

Data management tools are only as good as the underlying information and must be regularly updated to ensure that they continue to deliver relevant and accurate information to users (EBM Network 2009). Unfortunately, many tools suffer from a 'Cinderella effect', in which the user interfaces where data is presented often gain high profile and credibility, even while the underlying data upon which they draw are expected to maintain themselves or left to decay after an initial flurry of activity. The World Database of Protected Areas (WDPA), a joint initiative of IUCN and UNEP–WCMC, represents a key underlying data source, with supporting processes for compiling, verifying and updating information on legally protected areas from all nations. The Proteus Partnership was established precisely in order to ensure the long-term maintenance of the WDPA. The business community itself is an important collector and maintainer of BES data. ECOiSHARE is one attempt to create a platform for businesses to compile and share data with peers and others.

Sound science is critical for developing and maintaining data management tools. To date, most BES tools have focused on compiling and sharing spatial information on priority sites, based on species, habitats and ecosystems of 'global importance'.

Spatial data on the economic values of BES is still very scarce and there is a need for much more (and more consistent) valuations to enrich the information available for different sites and ecosystems. Some online libraries and websites have been developed to facilitate access to site-specific valuation case studies, for example the Environmental Valuation Reference Inventory (www.evri.ca) and the Conservation Value Map (www.consvalmap.org), but the wide variation in scales, objectives and methods used in different case studies means that comparisons across regions or landscapes can be difficult.

4.3.3 Modelling and scenario building

Several tools have been developed to facilitate modelling of potential future scenarios and risks, with an emphasis on ecosystem valuation. While these models may be seen as a subset of data management tools, their increasing prominence in detailed risk assessment and mitigation planning may be seen as a further step in the risk assessment process.

In most cases, such tools rely heavily on expert input, rather than simple automated processes to understand and interpret results. For example, the Integrated Valuation of Ecosystem Services and Tradeoffs (InVEST) is a scenario-modelling tool aimed at a broad audience, including business. It draws on a combination of pre-loaded and user-defined datasets to model the distribution of ecosystem services across areas of interest. The tool currently exists as an add-on to the popular ESRI Geographic Information Systems (GIS) software package. Similarly, the ARtificial Intelligence for Ecosystem Services (ARIES), currently under development, is aimed at a broad set of users that include business (Box 4.6).

Box 4.6 ARtificial Intelligence for Ecosystem Services (ARIES)

ARIES is a web-based decision-support tool that can be used to assess ecosystem services and model potential future scenarios. ARIES calculates the extent to which an area provides or uses a service, and how the benefits of ecosystem services flow across the landscape to reach beneficiaries. ARIES offers various entry points for decision makers and planners, including spatial assessments and economic valuations of ecosystem services, optimization of payment schemes for ecosystem services, and spatial policy planning. Different ecosystem services can be assessed at once, identifying areas that provide multiple 'bundled' services and trade-offs between them. An optional biodiversity layer integrates assessments of biodiversity and ecosystems services, allowing users to assess the value of biodiversity-rich areas and policy options that will ensure provision of ecosystem services.

In ARIES, the economic value of ecosystem services is a function of the biophysical flow of a given benefit that is provided to beneficiaries. Such flow information is used to mediate known values expressing demand from the users and market or non-market values derived from established literature sources made available through an internal economic database. Economic values are therefore a function of provision and usage, reflecting the ecological dynamics that determine supply on the one hand and the socio-economic drivers that determine demand on the other. In cases where ecosystem services are scarce or threatened, ARIES users can select non-linear functions to reflect potential changes in economic values in the vicinity of ecological thresholds.

The value transfer engine in ARIES then uses the quantitative assessment of flows and estimates of critical thresholds to apply values from existing studies to an area under investigation.

Recent applications of ARIES include designing payments for ecosystem services in biodiversity-rich areas of Madagascar, and assessing the benefits and beneficiaries of clean water in the cloud forest ecosystems of Mexico and of flood mitigation in Washington State, USA. Modelling the dynamic flow of benefits between terrestrial, coastal and marine ecosystems is another core activity in the ARIES project.

Source: Villa et al. (2009)

4.3.4 Economic valuation in decision making on risk

As described above, many tools have been developed to identify and manage the business risks associated with biodiversity, ecosystems and ecosystem services. However, with a few notable exceptions, these tools do not explicitly value BES in economic terms, for several reasons:

1 Most existing standards, guidelines and frameworks need reliable data, qualitative or quantitative, in order to generate useful outputs. There is, however, a lack of available data and interpretative guides to support higher-level analytical tools, resulting in dependence on project-specific assessment by experts. While non-economic assessments of BES also require fine-scale, site-specific analysis, there are more 'ready-made' tools that can present reasonably scaled bio-physical data at early stages of the project cycle (e.g. IBAT). All information collection comes at a cost; the challenge is to understand the trade-offs between investing in more information for decision making versus the costs of making bad decisions (Box 4.7).

Box 4.7 Investing in knowledge versus the costs of bad decisions: The case of water

The costs of managing water resources, including the protection of environmental values, are typically imposed on users through water charges. Such charges are necessary to cover the costs of ensuring long-term sustainability and security of access for water users. Some important considerations in managing risks to river health and groundwater-dependent ecosystems include:

- The risk of imposing overly conservative allocation rules, due to lack of information on environmental requirements, resulting in foregone consumptive use values.
- The opposite risk of over-allocating water to consumptive uses, resulting in loss of important ecosystem values.
- The costs of assessing ecosystem values accurately, monitoring the status of water resources and compliance with relevant regulations.

- The fact that rehabilitating degraded aquatic ecosystems generally costs more than resource protection (i.e. avoiding damage).
- Understanding the limits (thresholds) of ecosystem resilience and the irreversibility of certain environmental impacts.
- The potential role of water pricing and markets in allocating scarce resources to high value uses and promoting greater efficiency.
- Cost-sharing principles for investment in programmes to restore ecosystems, including user pays, beneficiary pays and historical legacy/liability.
- The importance of defining water resource allocations in law, in order to protect ecosystem values.
- Balancing environmental and socio-economic objectives, based on participatory policy formulation and sound science.

Source: Sutherland (2009)

2 The science of measuring biodiversity and ecosystem service values in economic terms has recently begun to be synthesized and communicated to business but has not been standardized or widely adopted (WBCSD 2011). In contrast, best practices for environmental impact assessment (EIA) have been distilled and standardized, drawing on a wide range of approaches to surveying and assessing biodiversity and ecosystem services. This has led to the development of a specialized sector of expert consultants, distinct from academic researchers, who are able to satisfy business needs (and statutory requirements) for impact assessment. A similar process of standardization and capacity development may emerge for economic valuation of biodiversity and ecosystem services, in response to business demand and/or changing regulatory requirements. The challenge for business is to define their precise economic valuation needs and for scientists to distil the minimum set of tools required to meet those needs.

3 Much of the discussion and development of biodiversity risk assessment and valuation tools for business has not benefited from systematic peer review. Most available information exists in the 'grey' literature or is linked to NGO or business activities with little or no input from other practitioners. As a result, trends, needs, tools and capacities for the valuation of biodiversity and ecosystem risks remain poorly documented. However, as tools for biodiversity assessment in business are further developed and applied, lessons will begin to emerge, and should benefit the broader community of users and developers.

4.4 Strategies for scaling down biodiversity and ecosystem risk

Best practices and assessment tools are important components of business risk management, but are not able to address all biodiversity risks. Risk management systems are generally designed to identify the issues that present the highest risks and to develop plans to avoid or mitigate these, while continuing to monitor other low-level risks. Effective monitoring may include strategies such as stakeholder engagement, partnerships and adaptive management to ensure that small risks do not become larger over time.

4.4.1 Stakeholder engagement

Many companies have developed community outreach and stakeholder engagement programmes, which are often managed separately from corporate policies and systems related to biodiversity and ecosystem services (see also Chapter 6 in this book). Stakeholder engagement, however, is an important part of managing biodiversity risks, as it can help identify potential threats and develop acceptable strategies for avoiding or mitigating risks, and monitoring changes over time. Stakeholders include anyone with the ability to affect the outcome of a project or the success of the business, whether positively or negatively, and can range from local communities and organizations to national governments and shareholders (IFC 2007). Stakeholder engagement is a complex process that should, at a minimum, include the following elements (IFC 2007):

- stakeholder identification and analysis;
- information disclosure;
- stakeholder consultation;
- negotiation and partnerships;
- grievance management;
- stakeholder involvement in project monitoring;
- reporting to stakeholders; and
- management functions.

Integrating biodiversity considerations into corporate stakeholder engagement policies and strategies can help mitigate risk. Relevant issues that may be considered as part of stakeholder engagement include (EBI 2003):

- local knowledge and use of biodiversity, including the rights of indigenous peoples and customary uses;
- local community dependence on natural resources for the supply of food, water, or other livelihood benefits; and
- aesthetic values and risks to human health.

Investing early in these processes allows companies to understand the local context, the values that local communities and other stakeholders place on biodiversity, and the extent of their reliance on ecosystem services that may be affected by business operations. Continuous engagement with a range of stakeholders over the lifetime of a business or project can also help in monitoring risks.

National and local governments are also key stakeholders for business engagement, as they regulate business access to or uses of biodiversity and ecosystem services. Engaging with government at all stages of a project can help identify impending changes in regulations, and how they might affect the company, and also present an opportunity to provide business input to regulatory reform processes.

4.4.2 Partnerships

Stakeholder engagement strategies can lead to the establishment of more formal collaboration or partnerships with NGOs and government, either on a bilateral basis or through multi-stakeholder processes. Such partnerships can be effective means for companies to manage biodiversity-related risks and may deliver other benefits as well (Table 4.2). Effective partnerships begin with a shared understanding of the issue(s) to

Table 4.2 Benefits of corporate–NGO partnerships

For the company	For the NGO
Enhances corporate reputation	Contributes to organization mission in new ways
Increases access to land and licence to operate	Increases access to new locations and networks
Helps to mitigate risk	Leads to involvement in integrated approach across a wider range of activities
Provides access to specialist expertise	Secures financial support for projects
Improves capacity to work with communities and access local information	Improves capacity for research, training and education
Builds corporate values and capacity of staff	Builds capacity of individual staff and institutions
Increases credibility with key stakeholders and leverage with other NGOs	Increases credibility and leverage with other corporations
Presents new opportunities to engage with external stakeholders	Builds innovative approaches to priority issues

Source: Adapted from Hurrell and Tennyson (2006)

be addressed and a shared commitment to objectives. They also require agreement on the expertise, resources and governance arrangements needed to undertake the work and to ensure credible results (Box 4.8).

Box 4.8 Partnerships in the building materials sector: a case study of IUCN–Holcim

In its first phase (2007–2010), the IUCN–Holcim partnership aimed to develop robust ecosystem conservation standards for the company, which could also contribute to sector-wide improvements in the cement and related sectors. The relationship was structured around three strategic objectives:

1 develop more comprehensive corporate biodiversity policy and strategy;
2 support sustainable livelihoods and biodiversity conservation through joint initiatives of mutual interest; and
3 promote good practice by sharing the lessons learned from the partnership with the wider industry and other stakeholders.

To support the partnership, IUCN established an independent panel composed of five experts in distinct but related fields. The panel was asked to provide input on biodiversity policy to the Holcim Group, to review existing biodiversity management tools used by the company and to suggest how these could be strengthened to conserve biodiversity more effectively, notably by bringing them together into an integrated biodiversity management system that would address all stages of the life cycle of a quarry, from initial scoping to closure. Members of the panel visited quarry sites at different stages of development (from 'greenfield' to closure) in Belgium, China, Hungary, Indonesia, Spain, the UK and the USA. Among other things, the visits allowed panel members to fine-tune their recommendations to the realities faced by Holcim in their everyday operations in different economic and cultural contexts.

The panel's independence was one of the most important aspects of the IUCN–Holcim relationship. It was agreed by all parties that the expert panel would only act as an advisory body, would not provide evaluations or assessments and would be accountable only to IUCN. Moreover, all final products generated by the expert panel would be made publicly available.

Building on the initial results of their collaboration, IUCN and Holcim agreed to renew their partnership over the period 2011 to 2013, with a focus on four areas:

1 Implementation of the biodiversity management system developed during the first phase, including new work on tools, capacity-building and indicators to ensure effective implementation of the system in Holcim's operations on the ground.
2 Policy influencing through joint work with policy makers to enable the building materials sector to deliver better biodiversity outcomes.
3 Sector-wide engagement, notably to influence the development of standards for biodiversity conservation in the building materials sector.
4 Collaboration to strengthen Holcim's approach to water management, including developing or adapting tools for measuring and mitigating water risks.

Sources: Carbone and Bos (2009a); Borges and Tong (2011)

Partnerships between conservation organizations and companies have emerged recently, in an effort to manage business impacts on biodiversity and ecosystems. Large environmental NGOs are increasingly leading multi-stakeholder initiatives to develop standards of good practice. In the agricultural sector, for example, WWF has joined corporate and other NGO partners to convene roundtables on sustainable palm oil and responsible soy, as part of wider efforts to develop standards of best practice for commodity production. At a more general level, several companies have joined efforts to foster business engagement in the Convention on Biological Diversity (Box 4.9).

Box 4.9 Business engagement initiatives under the Convention on Biological Diversity

The Convention on Biological Diversity (CBD) has recognized the importance of business engagement through a series of decisions, beginning in 2006 (see VIII/17, IX/26 and X/21). In 2008, at the ninth meeting of the CBD Conference of the Parties (COP-9), the government of Germany launched an initiative to promote business engagement. Over 40 companies committed to a Leadership Declaration on biodiversity, which was reinforced by the Jakarta Charter on Business and Biodiversity, adopted in November 2009.

In May 2010, in an effort to scale up and broaden business engagement towards COP-10 and beyond, the Japanese business community (including the Nippon Keidanren Committee on Nature Conservation, the Japan Chamber of Commerce and Industry and the Japan Association of Corporate Executives), in cooperation with IUCN and various government agencies (such as the Ministry of Agriculture, Forestry and Fisheries, Ministry of Economy, Trade and Industry and Ministry of Environment), announced a multi-stakeholder initiative called Japan Business and Biodiversity Partnership. The partnership was formally launched at COP-10 and brings together approximately 400 companies and business organizations, which have endorsed a Declaration on Biodiversity and have committed to promote actions to achieve CBD objectives, including sharing experiences and cooperation with NGOs, research institutions and governmental organizations.

Source: Furuta (2010), www.bd-partner.org

Assessing the benefits of multi-stakeholder partnerships both to business and for biodiversity can be challenging. Far too few partnerships define a clear baseline or develop a long-term monitoring framework that can assess not only changes in market conditions over time, but also the conservation, societal and business impacts of a partnership.

As consideration of biodiversity and ecosystem services becomes more of a norm within particular sectors, the task of coordinating multi-stakeholder initiatives may pass to industry associations. In the oil and gas and mining and metals sectors, for example, the Energy and Biodiversity Initiative (EBI), the International Petroleum Industry Environmental Conservation Association (IPIECA) and the International Council on Mining and Metals (ICMM) have all established working groups and/or developed best practice guidance on biodiversity for their members. However, such voluntary initiatives have generally not included external certification or verification and it is therefore often difficult to assess the level or quality of uptake of guidance issued on biodiversity by member companies.

Other examples of formal partnerships with external stakeholders have involved bilateral arrangements between a company and one or more NGOs to develop policies and management standards that integrate biodiversity and ecosystems considerations, or to understand the issues in relation to a particular site or sites. Rio Tinto (Hurrell and Tennyson 2006) and British American Tobacco (Box 4.10) offer examples of such partnership models.

Box 4.10 Integrating biodiversity into business: The BAT Biodiversity Partnership

From an agricultural perspective, tobacco is similar to other crops in its ecological impacts, with the difference that many farmers burn wood to cure tobacco leaves. While British American Tobacco (BAT) promotes tree planting wherever wood is used for curing, a proportion of wood is often still harvested from natural forests. Fast-growing plantations are part of the solution, but do not provide the same ecosystem services as native forests. The BAT Biodiversity Partnership, which includes Fauna & Flora

International, the Tropical Biology Association and Earthwatch Institute, was established in response to such issues.

The partnership issued a Biodiversity Statement in 2006, which recognizes the company's impact and dependence on biodiversity, as well as the need to assess impacts, engage with stakeholders, develop action plans and share information with suppliers. This in turn has led to the following commitments:

1 reduce to below 3 per cent the share of curing wood sourced from natural forest by 2015;
2 train managers, farmers and staff in biodiversity management; and
3 manage impacts in high-risk biodiversity locations through risk assessments and action plans.

To support this work, global biodiversity risk mapping as well as biodiversity risk and opportunity assessment and corrective action planning (BROA/CAP) tools have been developed. Such efforts have contributed to business sustainability through:

- ecological security and sustainability of agricultural/forest supply chains;
- securing the corporate licence to operate and meeting tighter regulations;
- maintaining reputation and setting higher standards in the sector;
- building stakeholder and shareholder trust and confidence;
- building employee support and strengthening the talent pipeline;
- finding better ways of working; and
- fostering collaborative local partnerships.

Source: Gilleard (2010)

Some bilateral NGO–business partnerships have started with broadly defined staff secondments intended to build trust and/or capacity (e.g. Shell and IUCN). Such arrangements can lead to deeper engagements that help companies identify and reduce biodiversity risks. For example, an Independent Scientific Review Panel (ISRP) was convened by IUCN from 2004–2005, at the request of the Sakhalin Energy Investment Company (a joint venture involving Shell International), to assess the risks posed by a major oil and gas project in the Sea of Okhotsk, off Sakhalin Island, and in particular the impacts on the western gray whale (www.iucn.org/wgwap/).

The ISRP experience generated useful lessons for how companies can work with and learn from conservation organizations to assess and mitigate biodiversity risks, for example the potential of independent reviews to:

- stimulate improved corporate biodiversity performance, while also increasing mutual understanding and respect among business and environmental organizations; and
- reduce scientific uncertainty and foster consensus on environmental solutions, even if definitive conclusions and universal agreement remain elusive (Halle with Hill 2009).

The ISRP experience also highlights the importance of transparency, independence and rapid follow-up when assessing corporate biodiversity risks, especially where the issues at stake are emotive or controversial.

Another way in which many companies address biodiversity risks is by developing partnerships with local communities around their facilities and operations. These can develop out of stakeholder engagement processes and may involve partnerships with NGOs to facilitate relationships. While these types of engagements often require significant investment by the company in stakeholder outreach, they can make a significant contribution to securing and maintaining the social 'licence to operate' in a particular region. Such engagements may include explicit payments for ecosystem services to ensure the continued provision of key ecosystem services on which the company depends, while also providing benefits, either financial or in kind, to local communities. For example:

- Vittel paid farmers in a watershed to adopt more sustainable land-use practices and restore ecosystems around water sources, in order to reduce the risk of water contamination (Perrot-Maître 2006).
- Energía Global supported a forest protection fund that pays landowners upstream of its dams to conserve or restore tree cover in order to reduce run-off and siltation (Porras and Neves 2006).

Not all biodiversity and ecosystem service related risks and opportunities can be successfully addressed through site-level regulations. Government agencies can help establish voluntary initiatives that provide compelling incentives for corporate participation in regulatory reform, reporting initiatives or continuous improvement.

4.4.3 Adaptive management

As companies adopt more rigorous environmental management systems and increasingly monitor and report their performance on biodiversity and ecosystem services, adaptive management approaches have proved to be extremely useful, especially given the lack of reliable information at the beginning of many projects. Adaptive management is a means of integrating monitoring and evaluation within a management framework that values learning from experience and facilitates the integration of this learning into improved management practices. It is an essential element in the continuous improvement of business operations.

The generic adaptive management framework consists of five steps, from (1) the conceptualization phase of a project through to (2) planning, (3) implementing action and monitoring results, to (4) analysing this information and adapting as necessary, and finally (5) sharing the findings more broadly (see Figure 4.4).

Adaptive management closely mirrors the ISO 14001 cycle, which follows a plan–do–check–act process. The key step of checking progress on a regular basis during the cycle to enable course corrections is a key component of adaptive management.

Within an adaptive management approach, monitoring and evaluation is built into every step of the process and is not left (or neglected) as a final step that is only sometimes included in a site management plan or corporate strategy.

The CBD has highlighted the importance of adaptive management for tourism development projects in vulnerable areas, in particular, due to lack of knowledge on these ecosystems and their biodiversity (SCBD 2004). Such advice could however extend beyond the tourism sector to include all industries with significant potential impacts on biodiversity and ecosystem services. This is especially important as

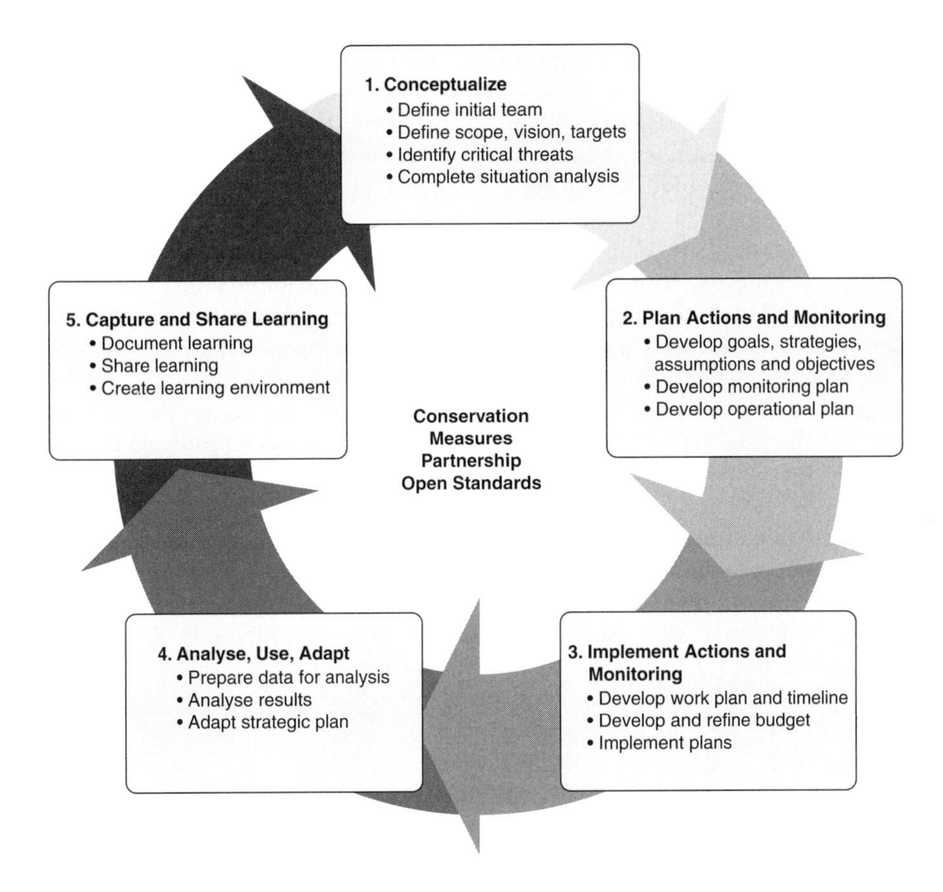

Figure 4.4 *Adaptive management framework*

Source: The Conservation Measures Partnership (2007)

companies consider the risks of climate change and develop strategies for adapting to an uncertain future (MacIver and Wheaton 2003).

Many corporate investments are managed over the long term and are subject to a high degree of variability over the investment life cycle. In addition to the direct impacts of the project itself, there may be external pressures that warrant careful monitoring and mitigation. Adaptive management systems provide a framework in which such discoveries and planning can take place. Addressing these issues, however, may require collaboration with other stakeholders, especially where external pressures are adding to ecosystem risks. Adaptive management should therefore look 'beyond the fenceline' of a project to identify and respond to such external risks at an early stage.

4.5 Conclusion

Businesses should broaden their risk assessment and management practices to reflect the increasing materiality of biodiversity and ecosystem services. Biodiversity loss and the decline of ecosystem services constitute material risks for companies – through shortages of key supplies, advocacy campaigns, fines and regulatory pressure, consumer boycotts or increasing difficulty in securing financing. These risks can be managed. A growing number of companies have discovered that it is feasible and cost-effective to integrate biodiversity and ecosystem services in their business risk management frameworks.

Businesses should develop internal capacity to understand and use existing biodiversity standards, frameworks and methodologies, data management tools, as well as modelling and scenario-building tools. A large menu of tools is available to help companies identify potential risks related to biodiversity and ecosystem services, to assess the degree of risk and to develop effective management strategies to mitigate the greatest risks. There is considerable potential to develop new tools, building on the established standards, frameworks and methodologies developed for specific sectors, across sectors or for use at various points along the value chain.

Economic valuation of biodiversity and ecosystem services can improve business decisions around risk management. Economic valuation of biodiversity and ecosystem services is a key tool for raising awareness of the magnitude of potential risks and the benefits of mitigation. More widespread use of valuation methods is feasible and desirable but is limited by several challenges, which, however, companies can help address. These include the need for more standardized approaches, and increased availability of data and capacity building on how to use valuation results in business decision making.

Businesses should actively pursue engagements and partnerships with a range of stakeholders. Companies are increasingly pursuing stakeholder engagement strategies as well as partnerships – with industry peers, NGOs, local communities, academia and others – to manage their biodiversity and ecosystem risks. Such engagements need to be more consistently documented in order to assess their effectiveness.

Acknowledgements

David Bresch (Swiss Re), Jürg Busenhart (Swiss Re), Toby Croucher (Repsol/IPIECA), Andrea Debbane (Airbus), Andrew Deutsch (Philips), Anne-Marie Fleury (ICMM), Juan Gonzalez-Valero (Syngenta), Paul Hohnen, Mira Inbar (Dow), Sachin Kapila (Shell), Chris Perceval (WRI), David Richards, Oliver Schelske (Swiss Re), James Spurgeon (ERM), Virpi Stucki (IUCN), Geanne van Arkel (InterfaceFlor), Mark Weick (Dow), Bernd Wilke (Swiss Re).

Notes

1 These include the ASX Corporate Governance Principles and Recommendations, the Combined Code on Corporate Governance, the King Code of Governance for South Africa, the US Sarbanes-Oxley Act and the Draft International Standard ISO 26000 (Guidance on Social Responsibility), respectively.

2 As seen, for example, in 'Beyond Grey Pinstripes', a survey and ranking of business schools focusing on how well they integrate social and environmental issues into curricula and research, published biennially by the Aspen Institute Center for Business Education (www.beyondgreypinstripes.org). In terms of course material, Harvard Business School has, for instance, published cases on the CBD and business engagement (Bell and Shelman 2006) as well as on farming and biodiversity (Lee and van Sice 2010). HEC Montréal has also produced a case on CDC Biodiversité, a subsidiary of the financial institution Caisse des Dépôts et Consignations (Wilain de Leymarie et al. forthcoming).

References

Austin, D. and Sauer, A. (2002) *Changing Oil: Emerging Environmental Risks and Shareholder Value in the Oil and Gas Industry*, World Resources Institute, Washington, DC. URL: www.wri.org/publication/changing-oil

BBOP (2009a) *Biodiversity Offset Design Handbook*, Business and Biodiversity Offsets Programme, Washington, DC. URL: http://bbop.forest-trends.org/guidelines/odh.pdf

BBOP (2009b) *Business, Biodiversity Offsets and BBOP: An Overview, Business and Biodiversity Offsets Programme*, Washington, DC. URL: www.forest-trends.org/biodiversityoffsetprogram/guidelines/overview.pdf

Bell, D.E. and Shelman, M. (2006) 'The Convention on Biological Diversity: Engaging the private sector', Case N9-507-020, Harvard Business School, Boston.

Borges, M.A. and Tong, R. (2011) Case material for TEEB, IUCN and Holcim.

Busenhart, J., Bresch, D., Wilke, B. and Schelske, O. (2010) Case material prepared for TEEB, Swiss Re, Zurich.

Carbone, G. and Bos, G. (2009a) Case material for TEEB, IUCN and Holcim.

Carbone, G., Tong, R. and Bos, G. (2009b) Case material for TEEB, IUCN and Holcim.

Conservation International (2010) *Integrated Biodiversity Assessment Tool for Business*, URL: www.ibatforbusiness.org.

Conservation Measures Partnership (CMP) (2007) *Open Standards for the Practice of Conservation* (Version 2.0). URL: http://conservationmeasures.org/CMP/Site_Docs/CMP_Open_Standards_Version_2.0.pdf

Earthwatch (2002) *Business and Biodiversity: A Guide to UK-based Companies Operating Internationally*, Earthwatch, Oxford. URL: www.businessandbiodiversity.org/pdf/BandBOseas.pdf.

EBI (2003) *Integrating Biodiversity Conservation into Oil and Gas Development*, Conservation International, Washington, DC. URL: www.theebi.org/pdfs/ebi_report.pdf

EBM Network (2009) *Using Ecosystem-Based Management Tools Effectively*, Ecosystem-Based Management Network. URL: www.ebmtools.org/sites/natureserve/files/Using%20EBM%20Tools%20Effectively_0.pdf

Evison, W. and Knight, C. (2010) *Biodiversity and Business Risk: A Global Risks Network Briefing*, World Economic Forum (WEF), Geneva. URL: www.weforum.org/pdf/globalrisk/Biodiversityandbusinessrisk.pdf

F&C Investments (2004) *Is Biodiversity a Material Risk for Companies? An Assessment of the Exposure of FTSE Sectors to Biodiversity Risk*, F&C Management Limited, London. URL: www.businessandbiodiversity.org/pdf/FC%20Biodiversity%20Report%20FINAL.pdf.

Furuta, N. (2010) Case material for TEEB, IUCN, Tokyo.

Gilleard, M. (2010) Case material for TEEB, Earthwatch, Oxford.

Halle, M. with Hill, C. (2009) *Learning for the Future: Lessons Learned and Documentation of the Process of Independent Scientific Review Panel for Western Gray Whales in Sakhalin*, IUCN, Gland, Switzerland.

Hanson, C., Ranganathan, J., Iceland, C. and Finisdore, J. (2008) *The Corporate Ecosystem Services Review: Guidelines for Identifying Business Risks and Opportunities Arising from Ecosystem Change*, WRI, WBCSD and Meridian Institute, Washington, DC. URL: http://pdf.wri.org/corporate_ecosystem_services_review.pdf

Holcim – Bos, G. and Tong, R. (2010) *Making Biodiversity Part of Business*, Holcim.

Houdet, J. (2008) *Integrating Biodiversity into Business Strategies: The Biodiversity Accountability Framework*, Fondation pour la Recherche sur la Biodiversité (FRB) and Orée, Paris. URL: www.fondationbiodiversite.fr/images/stories/telechargement/Guide-oree-frb-en.pdf

Hurrell, S. and Tennyson, R. (2006) *Rio Tinto: Tackling the Cross-sector Partnership Challenge*, International Business Leaders Forum, London. URL: www.thepartneringinitiative.org/docs/tpi/RioTinto.pdf

IFC (2007) *Stakeholder Engagement: A Good Practice Handbook for Companies Doing Business in Emerging Markets*, IFC, Washington, DC. URL: www.ifc.org/ifcext/enviro.nsf/AttachmentsByTitle/p_StakeholderEngagement_Full/$FILE/IFC_StakeholderEngagement.pdf

IPIECA and OGP (2005) *A Guide to Developing Biodiversity Action Plans for the Oil and Gas Sector*, IPIECA/OGP, London. URL: www.ipieca.org/activities/biodiversity/downloads/publications/baps.pdf

ISEAL (2010) *Code of Good Practice for Setting Social and Environmental Standards* (P005 – Version 5.01 – April), ISEAL Alliance, London. URL: www.isealalliance.org/resources/p005-iseal-code-good-practice-setting-social-andenvironmental-standards-v50

ISO (2004) *Environmental Management Systems: Requirements with Guidance for Use* (2nd edition), ISO, Geneva. URL: www.iso.org/iso/catalogue_detail?csnumber=31807

Johnson, S. (2006) *Good Practice Guidance for Mining and Biodiversity*, International Council on Mining and Metals (ICMM), London. URL: www.icmm.com/page/1182/good-practiceguidance-for-mining-and-biodiversity

Jones, M.E. and Sutherland, G. (1999) *Implementing Turnbull: A Boardroom Briefing*, Institute of Chartered Accountants in England and Wales. URL: www.icaew.com/index.cfm/route/120612/icaew_ga/pdf

Lee, D. and van Sice, S. (2010) 'Polyface: The farm of many faces,' Case 611001, Harvard Business School, Boston.

MacIver, D.C. and Wheaton, E. (2003) 'Forest biodiversity: Adapting to a changing climate,' Paper submitted to the XII World Forestry Congress, Québec City, Canada. URL: www.fao.org/DOCREP/ARTICLE/WFC/XII/0508-B3.HTM.

OECD (2005) *Environment and the OECD Guidelines for Multinational Enterprises: Corporate Tools and Approaches*, OECD, Paris. URL: www.oecd.org/document/36/0,3343,en_2649_34287_34992996_1_1_1_1,00.html

Perrot-Maître, D. (2006) *The Vittel Payments for Ecosystem Services: A 'Perfect' PES Case?* International Institute for Environment and Development, London. URL: www.iied.org/pubs/pdfs/G00388.pdf

Porras, I. and Neves, N. (2006) *Costa Rica – Energía Global: Energía Global Payments*, Central Plateau – watershed protection contracts, International Institute for Environment and Development (IIED), London. URL: www.watershedmarkets.org/documents/Costa_Rica_Energia_Global.pdf

PricewaterhouseCoopers (2010) *Biodiversity Offsets and the Mitigation Hierarchy: A Review of Current Application in the Banking Sector*, Study completed on behalf of BBOP and UNEP FI. URL: www.unepfi.org/fileadmin/documents/biodiversity_offsets.pdf

Rio Tinto (2008) *Rio Tinto and Biodiversity: Achieving Results on the Ground*. URL: www.riotinto.com/documents/ReportsPublications/RTBidoversitystrategyfinal.pdf

Sauer, A., Klop, P. and Agrawal, S. (2010) *Overheating: Financial Risks from Water Constraints on Power Generation in Asia*, World Resources Institute (WRI), Washington, DC. URL: www.wri.org/publication/over-heating-asia

SCBD (2004) *Guidelines on Biodiversity and Tourism Development: International Guidelines for Activities Related to Sustainable Tourism Development in Vulnerable Terrestrial, Marine and Coastal Ecosystems and Habitats of Major Importance for Biological Diversity and Protected Areas, including Fragile Riparian and Mountain Ecosystems*, SCBD, Montreal. URL: http://cdn.www.cbd.int/doc/publications/tougdl-en.pdf

Sutherland, P. (2009) Case material for TEEB.

Villa, F., Ceroni, M., Bagstad, K., Johnson, G. and Krivov, S. (2009) 'ARIES (ARtificial Intelligence for Ecosystem Services): A new tool for ecosystem services assessment, planning, and valuation', in *Proceedings of the 11th Annual BIOECON Conference on Economic Instruments to Enhance the Conservation and Sustainable Use of Biodiversity,* (September) Venice, Italy.

Waage, S., Stewart, E. and Armstrong, K. (2008) *Measuring Corporate Impact on Ecosystems: A Comprehensive Review of New Tools: Synthesis Report, Business for Social Responsibility (BSR)*. URL: www.bsr.org/reports/BSR_EMI_Tools_Application.pdf

WBCSD (2011) *Guide to Corporate Ecosystem Valuation: A Framework for Improving Corporate Decision Making*. World Business Council for Sustainable Development, Geneva, URL: www.wbcsd.org/web/cev.htm

Wilain de Leymarie, S., Raufflet E., Bertrand, N. and Brès, L. (forthcoming). *CDC (Caisse des Dépôts et Consignations) Biodiversité*. Centre de cas de HEC Montréal.

Chapter 5
Increasing Biodiversity Business Opportunities

Editors
Nicolas Bertrand (UNEP), Francis Vorhies (Earthmind)

Contributing authors
Robert Barrington (Transparency International UK), Joshua Bishop (IUCN), Ilana Cohen (Earthmind), William Evison (PwC), Lorena Jaramillo (UNCTAD), Chris Knight (PwC), Brooks Shaffer (Earthmind), Franziska Staubli (SIPPO), Jim Stephenson (PwC), Christopher Webb (PwC)

Contents

Key messages

Biodiversity and ecosystem services offer opportunities for all business sectors. The integration of biodiversity and ecosystem services into business can create tangible and significant added value for all companies and investors by increasing the cost-effectiveness of operations, ensuring the sustainability of value chains or enhancing revenues from new markets and new customers.

Biodiversity or ecosystem services can be the basis for new businesses. Conserving biodiversity and using biodiversity or ecosystem services sustainably and equitably can be the basis for unique business propositions, enabling entrepreneurs and investors to develop and scale up 'biodiversity businesses'.

Biodiversity and ecosystem service markets are emerging, alongside markets for carbon. New markets for biodiversity and ecosystem services are emerging – as have markets for reductions of CO_2 and other greenhouse gas emissions – providing new biodiversity assets with local and/or international trading opportunities. A key business opportunity is likely to be Reducing Emissions from Deforestation and Forest Degradation and related nature-based carbon storage and sequestration methods (REDD+).

Tools for building biodiversity business are in place or under development. Critical market-based tools for capturing biodiversity and ecosystem services opportunities – such as biodiversity performance standards for investors, biodiversity-related product certification, assessment and reporting schemes, and voluntary incentive measures – are already available or under development, and can be promoted across all business sectors and markets.

Appropriate policies can create an enabling framework for biodiversity and ecosystem service business opportunities. A range of voluntary and public policy measures at national, regional and international levels can create the enabling framework needed to scale up biodiversity and ecosystem services as viable business opportunities, such as payments for ecosystem services, REDD+, 'green development' certification, green tax incentives, biodiversity performance standards and development cooperation.

5.1 Introduction: Biodiversity as a business opportunity

How can businesses increase revenues, decrease costs and capture market shares by conserving biodiversity, restoring ecosystems or using biological resources sustainably? What new and promising business opportunities are available from conserving biodiversity, using biological resources sustainably and equitably, or supplying ecosystem services? This chapter examines these questions through the lens of three different approaches:

- By integrating biodiversity into business decision making, companies can enhance their performance by reducing risk, increasing revenues, reducing costs or improving their products.
- Biodiversity itself presents potentially huge untapped opportunities in the form of new goods and services (i.e. 'biodiversity business' opportunities).
- New markets for biodiversity and ecosystem services are emerging, inspired in part by the recent development of carbon markets. If scaled up, these markets could represent major business opportunities and a significant part of the solution to the biodiversity challenge.

Through a series of case studies and other examples, this chapter illustrates the diversity of realized or promising opportunities within these three categories. The chapter also considers the enabling conditions required to see these opportunities expand to their full potential. It defines a set of questions to help business identify such opportunities and identifies some of the potential challenges.

Table 5.1 provides a simple framework for identifying potential biodiversity and ecosystem service opportunities for various business sectors. It is similar to Table 4.1 in the previous chapter, which identifies biodiversity and ecosystem service risks to business. The key insight is that biodiversity and ecosystem service risks and opportunities are likely to be different in scope and significance for different sectors. The following section highlights a range of potential opportunities for several major business sectors, as indicated in Table 5.1.

5.2 Biodiversity and ecosystem services as a value proposition

'Biodiversity business' has been defined as 'commercial enterprise that generates profit via activities which conserve biodiversity, use biological resources sustainably, and share the benefits arising from this use equitably' (Bishop et al. 2008). Such businesses may focus on biodiversity-based production of commodities (food, timber, fabrics) or on the sustainable use of ecosystems (tourism, extractives, cosmetics, pharmaceuticals). In practice, any company that makes conserving biodiversity part of its core business can be considered a biodiversity business.

The number of businesses profiting directly or indirectly from biodiversity and ecosystem services has risen in recent years. Business leaders increasingly realize that biodiversity and ecosystem services offer new and attractive opportunities to improve their companies' profit margins and performance. Integrating biodiversity conservation and the supply of ecosystem services into business value chains or utilizing biodiversity and ecosystem services responsibly in production processes can

Table 5.1 Biodiversity and ecosystem service opportunities in different sectors

Categories	Indicative BES opportunities	Indicative market sectors with BES opportunities					
		Biological resource-based industries (e.g. forestry, farming, fishing)	**Extractive industries** (e.g. mining, oil and gas)	**Consumer goods** (e.g. clothing, cosmetics, furniture)	**Consumer services** (e.g. retailers, tourism)	**Health care** (e.g. pharma-ceuticals, biotherapy)	**Financials** (e.g. banking, biodiversity services, green investment funds)
Operational: day-to-day activities, expenditures and processes of the company	Increased quality, decreased cost of inputs	•		•	•	•	
	Increased output or productivity	•					•
	Sustainability of business operations	•	•				
	Supply chain opportunities			•	•	•	
Regulatory and legal: laws, policies, court actions that can affect performance	Lower transition costs in anticipating new policies		•				•
	Mitigation of risk due to environmental damage		•				•
Reputational: brand, image, relationship with stakeholders	Improved brand or image	•		•	•		
	Attract new customers			•	•	•	
	Reach new niche markets			•	•	•	
Markets and products: factors that can affect corporate market performance	Changes in consumer preferences	•			•	•	
	Purchaser requirement	•		•			
Financing: cost and availability of capital	Attract growing SRI investment	•	•	•	•	•	

Source: Adapted from Evison and Knight (2010) and Hanson et al. (2008)

result in significant savings and increased revenues for businesses. Prudent management of biodiversity and ecosystem service risks can deliver tangible savings, while biodiversity and ecosystem services can unleash hidden value in production and marketing practices. The benefits of integrating biodiversity and ecosystem services in business are especially advantageous in that they can help achieve corporate social responsibility objectives at a time when the general public is increasingly concerned about biodiversity loss and the importance of conservation.

Growing public concern about the environment, combined with consumers' increasing focus on health, wellness and humane sourcing, is changing the marketplace. For example, so-called 'lifestyles of health and sustainability' (LOHAS) consumers focus their consumption patterns on health and fitness, the environment, personal development, sustainable living and social justice (see www.lohas.com). The LOHAS market today includes over 80 million consumers, representing a potential demand of US$500 billion for organic food products, sustainable tourism, green building suppliers, low-energy appliances, and so on.

Within the LOHAS market, biodiversity and ecosystem services offer a range of business opportunities as highlighted in the following illustrations. The returns from biodiversity and ecosystem services emerge both directly from conservation and sustainable use, and indirectly through the integration of biodiversity and ecosystem services into existing product lines.

5.2.1 Agriculture

Many consumers today prefer organic foods. In addition to perceived health benefits, these consumers are also seeking traceability, ethical sourcing, sustainability, and corporate social and environmental responsibility (Organic Monitor 2009a). In response to these changing consumer preferences, several major brands are shifting some of their goods toward natural, fair trade, organic or sustainable labelling (Kline & Company 2009).[1]

Worldwide sales of organic food and drink in 2007 amounted to some US$46 billion, a threefold increase since 1999 (Organic Monitor 2009b). In the USA alone, organic food sales accounted for US$22.9 billion or 3.5 per cent of the total food market and grew by 15.8 per cent in 2008 or almost triple the growth rate of the food sector as a whole in the same year (OTA 2009).

The concept of 'fair trade' has also gained currency as a means to identify more sustainable agricultural products. In 2008, sales of fair trade certified products totalled US$4.08 billion worldwide, a 22 per cent increase on the previous year. Although this represents a tiny fraction of overall world trade, fair trade labelled products in some categories can account for up to 20–50 per cent of total sales. In June 2008, it was estimated that over 1 million farmers and their families were benefiting from fair-trade-funded infrastructure, technical assistance and community development projects (Fairtrade Labelling Organization 2010).

Other initiatives to promote more sustainable agricultural practices include the efforts of the Rainforest Alliance, which has certified over 205,000 hectares in the tropics for the production of bananas, cocoa, coffee, tea, citrus, flowers, pineapples and other commodities. These schemes cover more than one million producers in 14 countries and seek to ensure positive impacts on biodiversity and ecosystem services (Wille 2009).

The agriculture sector remains a major driver of land-use change with significant adverse impacts on biodiversity and ecosystem services. Efforts to promote more sustainable agricultural practices, including certification and labelling schemes such as those highlighted above, are a key means for the food industry to address its impacts on biodiversity and ecosystem services. Further examples of business action on biodiversity in the agriculture sector are provided in Boxes 5.1 and 5.2.

Box 5.1 Chocolats Halba: Ensuring cocoa bean security and partner satisfaction

Chocolate and confectionery company Chocolats Halba, a subsidiary of the Swiss retailer COOP, has integrated sustainable sourcing into its cocoa supply chain. As in the rest of the chocolate industry, the company faces frequent supply shortages. Because of price instability, cocoa production is more risky for small-scale farmers, who produce most of the world's cocoa; many of them are looking for alternative employment opportunities. Chocolats Halba discovered that the best way to support farmers was to establish diversified agroforestry systems that include cocoa as one of many crops. In such systems, biodiversity is generally higher because they establish a more species-diverse landscape. With agroforestry systems, farmers also have a better and more diverse income, which means that cocoa farming is not only good for biodiversity and ecosystem services, it is also good for making cocoa growing more attractive to farmers.

According to Christoph Inauen, head of Chocolats Halba Sustainability and Projects:

> *Farmers that work with us realize that we are not only interested in cocoa but also in their livelihood, their income, biodiversity (we help them reforest deforested areas), and other issues. This makes our relationship very strong: farmers give their best to improve the quality of the cocoa in order to give us something back. So we have reliable sourcing partners and very strong relationships with our farmers. In the case of a supply shortage, this would surely help us.*

Source: Inauen (2010a, b)

Box 5.2 Conservation Grade nature-friendly farming

The UK-based Conservation Grade certification system of 'nature-friendly farming' provides food brands, producers and consumers with a solution to efficient food production while enhancing biodiversity and ecosystem services and preventing wildlife declines on farmland. It does so by requiring Conservation Grade farmers to take 10 per cent of their land out of production for conversion to wildlife habitats. In return, these farmers are able to use the Conservation Grade logo on all their products and have access to a supply contract for their produce for which there is a guaranteed premium over the market price.

The Conservation Grade farming scheme is an example of an innovative solution to feeding a growing world population without destroying biodiversity. Independent trials have demonstrated that the Conservation Grade farming system leads to significant increases in biodiversity, as compared to conventional agriculture, while maintaining production and agricultural incomes.

Source: www.conservationgrade.org

5.2.2 Biodiversity management services

Biodiversity management and related services include advising companies on possible biodiversity and ecosystem service risks during an environmental impact assessment, developing corporate biodiversity strategies and action plans, and independent auditing or review of a company's biodiversity performance.

Although most biodiversity management services focus on mitigating biodiversity risks, there are also services that focus on how to capture biodiversity opportunities – for example, by exploring how biodiversity-friendly supply chains can improve a company's reputation among investors or consumers. One example is a recent collaboration between a large German energy company, Eon, and IUCN on so-called 'blue energy'. IUCN helped the company to understand both the biodiversity risks and opportunities of supplying electricity from offshore renewable technologies, such as wind farms, and thus enabled the company to manage its biodiversity impacts – both negative and positive (Wilhelmsson et al. 2010). By making offshore wind power more friendly to biodiversity, Eon hopes to strengthen its position with investors, consumers and government regulators.

5.2.3 Cosmetics

Natural cosmetic companies were among the early adopters of organic and fair trade certification schemes. They recognized the importance of sourcing their ingredients responsibly and launching certified products well before many other industries. Some of these companies have set up their own fair trade and organic projects to protect endangered plant species and to encourage the sustainability of wild harvesting. Others are using fair trade certification to guarantee the long-term supply of organic ingredients, while a dedicated sustainability standard for wild harvested products is also in the works (www.fairwild.org). Major companies in the organic cosmetic sector reported positive growth for 2008, including Weleda (with a 9.5 per cent increase in sales to €238.3 million), Wala (with 'Dr. Hauschka' brand sales increasing by 7.3 per cent to €103 million) and Lavera (with an increase of 16 per cent to €35 million; BioFach 2009a, b).[2]

5.2.4 Extractive industries

Oil and gas, mining and quarrying companies, by the very nature of their operations, can have significant impacts on biodiversity and ecosystem services. These impacts can and often do lead to negative consumer perceptions of these industries. Businesses in extractive industry sectors are therefore well placed to take advantage of the opportunities created by biodiversity and ecosystem services. Through the development and effective implementation of a corporate biodiversity action plan (BAP), an extractive company can reduce the adverse impacts of its operations on biodiversity, while also enhancing support for rehabilitation and conservation in the landscapes in which they operate. Two examples are provided in Boxes 5.3 and 5.4.

Box 5.3 Rio Tinto: Towards net positive impact on biodiversity in mining

The mining giant Rio Tinto aims to have a 'net positive impact' (NPI) on biodiversity. It has launched pilot schemes in Australia, Madagascar and the USA in partnership with conservation organizations such as Earthwatch Institute, FFI and IUCN. The company is also assessing the biodiversity values of its landholdings and developing site-level biodiversity action plans (BAPs). Some key outcomes of Rio Tinto's biodiversity work include:

- better understanding of how quantitative measures of biodiversity performance can be used to assess progress;
- increased awareness and knowledge of biodiversity through the active participation of company staff in ongoing field research programmes; and
- a concrete demonstration of how a business can work with non-profit civil society organizations to promote biodiversity conservation and sustainable management of ecosystems, benefiting both the company and the environment.

Source: 2009 Annual Report, Environmental Stewardship chapter: www.riotinto.com/annualreport2009/performance/sustainable_development_review/environmental_stewardship.html

Box 5.4 Yemen LNG: Investing in the protection of marine biodiversity

Yemen LNG Company Ltd is involved in a large-scale project that includes the construction and operation of a liquefied natural gas plant in the area of Balhaf, Republic of Yemen – a remote spot on the country's south coast. The company has implemented a marine biodiversity action plan in order to improve its environmental performance and in turn its attractiveness to investors and the public. A major challenge for such a project is arranging credible, independent verification of the implementation of environmental strategies.

In early 2009, Yemen LNG finalized a partnership with IUCN to design and manage an independent review process of the company's marine biodiversity action plan. Specifically, the agreement aims to secure independent third-party assessment of the company's strategy for marine biodiversity protection and the implementation of its biodiversity action plan. Through such an independent audit, the company hopes not only to improve its performance with respect to the environmental and social aspects of its marine biodiversity strategy, but also to be able to demonstrate to its board, investors and other stakeholders that it is doing so.

Source: http://yemenlng.com/ws/en/go.aspx?c=soc_Environment

5.2.5 Finance

The financial services sector stands to benefit from biodiversity and ecosystem services in many interesting ways, including improved stakeholder perception, streamlined operations, enhanced ability to attract talent, increased profit through investments in biodiversity and ecosystem services and, importantly, reduced risk for investors.

Several dedicated bio-enterprise investment funds have recently been established to invest in businesses that demonstrate a potential to deliver both financial returns and biodiversity benefits. Some funds combine financing with business development assistance and management capacity-building, as an integrated package. Examples include Verde Ventures, an arm of Conservation International (CI), the Eco-Enterprises Fund, an arm of The Nature Conservancy (TNC), and Root Capital, formerly EcoLogic Finance. At a broader scale, Box 5.5 provides an example of how a mainstream financial institution is exploring biodiversity opportunities.

Box 5.5 HSBC: Developing biodiversity awareness in a bank

In 2002, HSBC launched 'Investing in Nature', a US$50-million, five-year partnership with several conservation organizations. As part of this initiative, HSBC sent 2,000 of its employees on Earthwatch Institute field research projects around the world and supported the training of 230 developing-country scientists. Participating HSBC employees also had a responsibility to undertake an environmental project in their workplace or local community, supported by a small grant from the company. This initiative not only supported conservation on the ground but did so in a way that strengthened biodiversity awareness and commitment across the HSBC workforce.

Employees who participated in 'Investing in Nature', as well as those who take part in the more recent 'HSBC Climate Partnership', bring their newfound knowledge of biodiversity back to the company, acting as 'environmental champions' within the business and developing projects that further HSBC's commitment to sustainability. Such projects can also help increase the company's ability to attract young talent (Connor 2010).

A recent independent evaluation, undertaken by the Ashridge Business School, found that 80 per cent of senior HSBC managers agreed that the programme contributes to embedding sustainability in the business, while 83 per cent agreed that it is worth the investment of time required because the programme gives HSBC a competitive advantage.

Sources: www.hsbc.com/1/2/newsroom/news/2002/investing-in-nature-2002#top and www.earthwatch.org/europe/our_work/corporate/corporate_partners/hsbc/hcp

5.2.6 Fisheries

Due to the severe depletion of many wild fish stocks, fishing fleets and companies that buy their products are increasingly concerned about the sustainability of supply. The sector also has an opportunity to cater to the growing number of LOHAS consumers who want to consume fish – often considered a healthy alternative to meat – but worry about negative impacts on the environment. Several voluntary certification and eco-labelling schemes for sustainably managed fisheries have emerged, providing assurance to both business buyers and consumers. Perhaps the most widely recognized scheme for certifying sustainable seafood is the Marine Stewardship Council (MSC; www.msc.org).

Many companies in the sector, from suppliers to sushi restaurants, claim that carrying the MSC label has provided new business opportunities. These include access to new markets – both geographically and in terms of niche markets from sustainable product categories – and retention of existing markets. Small-scale fishermen report price premiums for their certified fish. For the US-based seafood restaurant Bamboo Sushi, selling only sustainable seafood is a lucrative business model (www.bamboosushipdx.com). Some big retailers, including discounters Aldi and Lidl, request MSC-certified frozen products from their suppliers. With consumers and seafood buyers increasingly aware of the importance of healthy oceans, the market for MSC-certified seafood has grown to an estimated US$2.5 billion (Marine Stewardship Council 2009).

5.2.7 Forestry

Growing consumer preference for wood and paper products that are certified as coming from sustainably managed forests has strongly driven this industry's recent development and marketing techniques. Forest certification under a recognized standard is an important requirement to access the EU market, which is one of the world's largest consumers of timber and related products. Certified timber is also increasingly available in many other markets.

Leading certification schemes for the forest sector include the Forest Stewardship Council (FSC), the Programme for the Endorsement of Forest Certification (PEFC) and the Rainforest Alliance (RA) SmartWood programme. By May 2009, the global area of certified forest endorsed by FSC and PEFC amounted to 325.2 million hectares, or approximately 8 per cent of global forest area. Certified wood products are now common in many large retailers like B&Q in the UK, Home Depot in the US and Ikea worldwide.

5.2.8 Garments

According to *Environmental Leader*, consumer preference for ethically sourced, organic and fair trade fabrics will affect the garment industry as well (Willan 2009). So-called 'eco-fabrics' are thought to add value to products (Prescott 2009), while natural fibres – mainly cotton and blends – are increasingly preferred by consumers over synthetic fibres (CBI 2008).

In the garment and textile sectors, organic cotton has become a key marketing tool for many companies. Today, organic cotton cultivation covers about 32 million hectares of farm land (FiBL and IFOAM 2009). Global retail sales of organic cotton clothing and home textiles totalled over US$3 billion in 2008 (www.naturalfibres2009.org).

Especially in Europe, there is a widespread demand for natural fibres in addition to organic cotton. Natural fibres such as pima cotton, alpaca wool or mohair wool have also recently become popular. They are mainly used in high-end products due to their relatively high production and raw material costs and therefore remain a niche market. Developing country companies such as the Star Knitwear Group of Mauritius presented fabrics at Texworld 2009 based not only on African cotton but also bamboo, corn, Tencel and Modal (Prescott 2009; www.eartheasy.com).

Sustainability is also an issue for leather used in the garment and accessories industries, for example jackets, belts, purses, luggage and wallets. Manufacturers increasingly demand skins from animals such as crocodiles, lizards and snakes that are sustainably managed and legally traded. There are opportunities for this sector to engage in profitable biodiversity conservation, as consumers insist on eco-friendly leather goods, whether made from recycled materials or by using environmentally friendly production processes such as tanning (CBI 2009b; Mazzanti and Zavettieri 2009).

5.2.9 Handicrafts

The handicrafts sector is another area that provides opportunities for biodiversity-friendly job creation, especially in developing countries. In Vietnam, for example, the production of handicrafts directly involves almost two thousand craft villages and is expected to generate US$1.5 billion in turnover by 2010 (VIETRADE n.d., 2008). The products are generally made from biological resources (bamboo, rattan, rush, leaf, wood, etc.) and other natural materials such as metal and stone.

Handicrafts are strongly influenced by fashion trends, consumer preferences and overall economic conditions (Barber and Krivoshlykova 2006). Social and environmental values are gaining prominence in this sector as well, with a fair trade movement appearing in the handicrafts and decoration sector. An international label for this sector is now available from the World Fair Trade Organization (WFTO; www.wfto.com).

5.2.10 Pharmaceuticals

Many industries use wild genetic resources as inputs to production. Over 400,000 tons of medicinal and aromatic plants are traded worldwide every year; 80 per cent are wild harvested, in most cases with little consideration of where they come from or the sustainability of collection practices (Traffic International 2006). Demand continues to grow for these valuable plants.

The pharmaceutical industry is an important user of wild genetic resources, along with biotechnology, seed producers, animal breeders, crop protection services, horticulture, cosmetics, fragrance, botanicals, and the food and beverage industries. Each sector is part of a unique market, with distinct research and development processes and different demands for access to genetic resources (Laird and Wynberg 2005).

Consistent data on the use and value of genetic resources do not exist for most sectors and only rough estimates are available. For example, ten Kate and Laird (1999) suggest that between 25 and 50 per cent of the value of global pharmaceutical sales is based on the use of genetic resources. Based on the total market value of the pharmaceutical sector – currently around US$825 billion, according to IMS

(2009) – these ratios, if still valid today, imply that the value of genetic resources in the sector might be up to US$412 billion. By comparison, the commercial global seed market, which also relies on wild genetic material, is expected to reach a total value of US$42 billion in 2010 (Global Industry Analysts 2008).

Wild ingredients and other raw materials used in the medical sector may be certified organic but are often not labelled as such, as the final product typically does not allow for this. More generally, producers can refer to the Good Agricultural and Collection Practices (GACP) for medicinal plants (World Health Organization 2003), to ensure that documentation is complete and traceable in order to guarantee the origin and quality of wild products. Adherence to such practices can help ensure a stable source of raw materials and thus a more secure supply chain, as well as reducing the risk of accusations of 'bio-piracy' or inadequate benefit sharing (for further discussion, see Chapter 6).

5.2.11 Retail

Large retailers can have significant positive impacts on biodiversity and ecosystem services, through such measures as sustainable sourcing, discernment in choosing which items to stock, and improved packaging and distribution techniques. In return, these companies can benefit from lower operating costs, improved customer loyalty and increased supply chain security. Increasingly, biodiversity responsibility is being mainstreamed into the retail sector, with companies such as COOP in Switzerland, Whole Foods in the USA, and Marks & Spencer in the UK capturing 'first mover' advantages. Box 5.6 provides an example of sustainable sourcing in the world's largest retail company, Walmart.

Box 5.6 Walmart: Stocking sustainable products to meet consumer demand

In 2005, Walmart announced a new environmental strategy involving, among other objectives, a commitment to sell more sustainable products. The company has developed a 'sustainable product index' to assess the environmental impacts of the goods it stocks and relays this information to customers using a labelling system. The sustainable product index measures such facets of production as energy usage, material efficiency and human conditions. Products with higher scores have lower environmental footprints and promote biodiversity conservation in various ways.

Walmart states that its customers want 'products that are more efficient, last longer and perform better. They want to know the product's entire lifecycle. They want to know the materials in the product are safe, that it is made well and is produced in a responsible way.' The company believes that its sustainable product initiative will increase customer numbers and business revenues, while at the same time reducing expenses – for example, through decreased dumpster costs from the elimination of excess product packaging.

Sources: Plambeck and Denend (2008); http://walmartstores.com/Sustainability/; Bernick and Guth (2010)

5.2.12 Tourism

Consumer trends in favour of environmentally sustainable activities have positively affected the tourism sector. One survey indicates that 'sustainable travellers' are willing to spend an average of 10 per cent more on travel services and products provided by environmentally responsible tourism suppliers (CBI 2009a).

Many travel agents have realized that sustainable tourism provides an excellent market opportunity, in which economic profit and respect for the environment go hand in hand. By 2009, in South Africa – where the first label for FairTrade Tourism was developed – about 45 tourism products had been certified, including hotels, safari lodges, guest houses, backpacker lodges, eco-adventure activities and township tours. Both local and international tour operators rely on certification according to the standards of FairTrade Tourism South Africa, including operators from Germany, Switzerland and the UK (CBI 2004, 2009c; Fair Trade in Tourism South Africa 2009).

More broadly, the International Ecotourism Society (TIES) has developed a global network to support 'responsible travel to natural areas that conserves the environment and improves the well-being of local people' (www.ecotourism.org). With members in over 90 countries, including more than 50 local, national and regional ecotourism associations, TIES engages in awareness raising and educational projects to build the market for ecotourism.

Further evidence of the growth of this sector was the announcement in 2009 that the Partnership for Global Sustainable Tourism Criteria (GSTC) and the Sustainable Tourism Stewardship Council (STSC) would merge to form a new Tourism Sustainability Council (TSC 2009), with the aim of promoting common understanding of sustainable tourism and the adoption of universal sustainable tourism principles and criteria.

Today there are many hundreds of examples of successful ecotourism companies and a growing number of organizations that promote these companies. For instance, the Athens-based EcoClub promotes 'ecologically and socially just tourism' (www.ecoclub.com), while Planeta.com offers a World Travel Directory 'for those seeking meaningful eco-friendly, people-friendly, and place-friendly travel' (www.planeta.com).

5.2.13 Biodiversity: An opportunity to scale up business

Even companies lacking an apparent direct interaction with the natural world can find opportunities and incentives in the increasing societal and consumer preferences to 'go green'. For example, companies like GreenGeeks provide environmentally friendly internet services for companies and individuals. Consumers increasingly value such products and services that have a positive and meaningful association with the environment. Corporate action on biodiversity and ecosystem services can help businesses distinguish themselves from their competitors, while also improving relations with investors, employees and communities. This shift can attract new investment capital – whether from socially conscious investors or the growing number of mainstream investors who recognize good environmental management as a proxy for overall sound business management. It can also help attract top young talent, who are increasingly aware of the fragility of the natural world and the potential impact of business – both negative and positive – on it. Corporate biodiversity responsibility also builds good will with stakeholders both locally and internationally.

There is already a diverse array of biodiversity business opportunities being captured in existing market sectors and segments, as outlined above. Further opportunities to scale up biodiversity businesses or to develop biodiversity-friendly products and services are likely to emerge in the near term. In addition, entrepreneurs and investors can help shape and profit from entirely new markets for biodiversity and ecosystem services, as described further below.

5.3 Emerging markets for biodiversity and ecosystem services

Ecosystem service markets may be defined as 'the bringing together of a buyer and seller so that they can trade ecosystem service credits' (Ecosystem Services Project 2008). According to this definition, ecosystem service credits can be considered as marketable units that represent the protection and/or enhancement of ecosystem services. Probably the most well-developed and widely known example of market creation for an ecosystem service is the global market for carbon credits (Box 5.7)

Box 5.7 The development of the carbon market: A model for biodiversity?

The global carbon market grew from virtually nothing in 2004 to reach a total value in excess of US$140 billion by 2009 (see chart below). This market included project-based transactions, such as Clean Development Mechanism (CDM) projects and the voluntary market, as well as the so-called 'allowances market', such as the EU Emissions Trading System (ETS). A key driver of the rapid growth in the carbon market was a commitment by several governments (especially in the EU) to reduce emissions of greenhouse gases, and the translation of this commitment into allocations of tradable emission permits to industry.

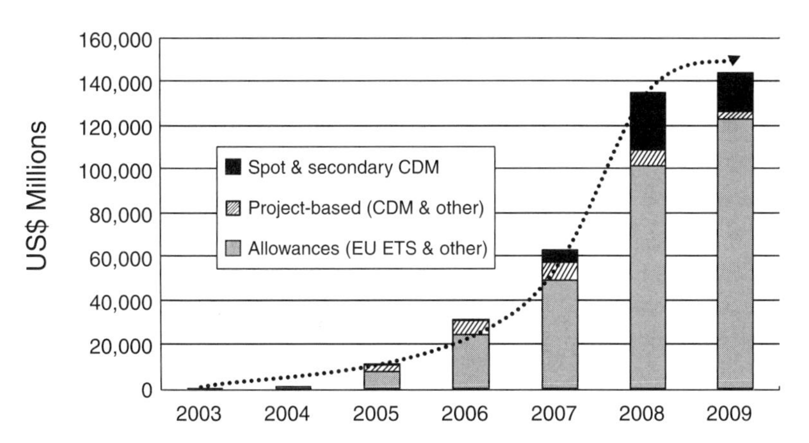

In parallel with the creation of the carbon market, which by itself made renewable energy more competitive, many governments established additional targets and incentives to promote renewable (low-carbon) energy generation. As a result, global revenues of companies involved in the wind, solar and bio-fuels markets reached US$116 billion in 2008 – almost a tenfold increase over five years. One sector alone – wind power – had revenues exceeding US$50 billion. The long-term forecast is for further growth of renewable energy supplies to over US$300 billion annually by 2020. One estimate, based on a broader definition of climate change-related sectors, including energy efficiency, waste and water, suggests that global revenues already exceed US$530 billion and could grow to US$2 trillion by 2020 – which would make the sector comparable in size to today's global oil and gas industry.

Investment is yet another indicator of how climate policy and the carbon market have changed the business landscape. According to a recent UNEP report, investments in sustainable energy technologies – including venture capital, project finance, public markets, and research and development – reached US$155.4 billion in 2008, despite the economic downturn.

Sources: Kossoy and Ambrosi (2010); Clean Edge (2009); HSBC (2009); UNEP and New Energy Finance (2009); http://ec.europa.eu/environment/climat/emission/index_en.htm

The establishment of market-traded credits is a key characteristic of the carbon market, and one that distinguishes it from some other environmental payment schemes, such as direct payments for ecosystem services (e.g. payments by Perrier-Vittel to farmers in France to reduce agricultural pollution, or by the New York City water authority to landowners in the Catskill/Delaware and Croton watersheds), or product certification schemes (e.g. Marine Stewardship Council).

Whatever mechanism is used, the common reason for establishing payments or markets for ecosystem services is to 'internalize' the external ecological costs of doing business (as opposed to using ecosystems 'for free', which tends to lead to overuse). Who ends up paying the cost of using ecosystem services will depend on market structure, the availability of substitutes and other factors. In general, we can expect that part of the cost will be absorbed by business and part may be passed on to the final consumer. Some reasons why many businesses are engaging proactively in ecosystem service markets are outlined in Table 5.2.

Table 5.2 The business case for engaging in ecosystem service markets

1. Market differentiation	Businesses that take a lead in ecosystem service markets can differentiate themselves from their competitors and are more likely to keep ahead of regulatory, investor and consumer expectations. Benefits may include increased regulatory goodwill, reputational gains, and potential improvements in staff recruitment and retention (Mulder et al. 2006), as well as helping to ensure the long-term sustainability of business operations.
2. Revenue generation: selling credits	The growth and diversification of ecosystem service markets is increasing the opportunities to invest in or develop projects that generate revenue through the sale of environmental credits. Recent trends suggest this growth will continue, although ambiguities around the pace and nature of future regulatory change mean that the pattern of growth of such markets in different jurisdictions and for different ecosystem services is very uncertain.
3. Revenue generation: ancillary services	The demand for technical assistance and other ancillary services linked to ecosystem markets is likely to grow in conjunction with the expansion and diversification of such markets. As competition increases, markets will benefit from higher support service standards, with early movers defining these standards to a great extent. It is these early movers who are likely to establish themselves in ecosystem service markets and to receive the bulk of the financial rewards.

5.3.1 Regulatory markets for biodiversity and ecosystem services

Markets for biodiversity and ecosystem services can be broadly divided into regulated or 'compliance' markets and voluntary markets. This section examines ecosystem service markets that are enabled by government regulations, while voluntary markets for ecosystem services are considered in the following section.

The simplest version of a regulated market for an ecosystem service is when the government or some other regulatory body imposes a limit or 'cap' on the degree of ecosystem use or damage that is legally allowed in a certain jurisdiction or creates tradable quotas for specified 'sustainable' activities (e.g. wind power generation). Typically, the regulator further allows firms or individuals to trade ecosystem service credits in order to meet their legal obligations, i.e. selling surplus credits or buying additional credits that are needed to match the firm's use of ecosystem services or its environmental footprint (Fischer 2003). These kinds of trades take place because the costs of reducing environmental damage or of meeting a quota are rarely equal for all companies, due to differences in technology or other factors. In general, firms for whom the costs of compliance are relatively high will tend to purchase additional credits from other firms that are able to meet their quota at lower cost or have more credits than they need for some other reason. Examples are provided in Box 5.8.

Box 5.8 Selected examples of regulated markets for ecosystem services

- **Australia – BushBroker:** The clearing of native vegetation in the state of Victoria is primarily regulated by the Victoria Planning and Environment Act of 1987. In 2006, the Victoria government introduced the BushBroker scheme on a pilot basis, which allowed for clearing of native vegetation only when compensated by a comparable in-kind offset.

  Permit applicants were able to source these offsets through the BushBroker register. Offsets are defined as gains in native vegetation extent and/or condition that are permanently protected and linked to a particular clearing site. Applicants can either generate offsets on their own property or purchase these offsets as native vegetation credits from third parties. Over AU$4 million worth of agreements were facilitated by the programme in its initial phase.

  The BushBroker scheme also allows for the 'banking' of credits for future use. For instance, a construction company could donate land to the conservation reserve system and register the resulting credits as offsets for its own future use. However, the main revenue opportunity for business is by generating native vegetation credits through improved land management, re-vegetation of previously cleared areas and protection of existing stands of trees. This relatively low-cost process can deliver significant income streams from land of low commercial value. Average prices for credits have ranged from AU$42,000 to AU$157,000 per hectare.

- **USA – Biodiversity banking:** The US Endangered Species Act prohibits any development that harms populations of endangered species. Landowners and developers are legally obliged to mitigate impacts on endangered species but may buy offsetting credits from approved biodiversity banks. Examples include the Mariner Vernal Pool Conservation Bank, a 160-acre bank that by early 2007 had sold US$4.4 million worth of credits; and the Sutter Basin Conservation Bank, a 424-acre bank set up to generate revenue by selling giant garter snake habitat credits. Equivalent to one acre of snake habitat each, these are purchased by developers in order to fulfil their legal obligations to the US Fish and Wildlife Service and the California Department of Fish and Game for protecting giant garter snake habitat.

- **USA – Wetland banking:** The federal Clean Water Act obliges developers to replace as many wetlands as they drain, fill or otherwise damage. Developers can purchase credits, which come in the form of wetland acres, within the same watershed as their development to offset ecological damages.

  Of the 450-plus approved wetland banks in the USA, at least 20 per cent were developed by large corporations, predominantly energy or pipeline companies such as Chevron, Tenneco, and Florida Power and Light. These companies have surplus land, are often looking for ways to diversify their income streams and are attracted by the relatively low costs of setting up wetland mitigation banks. For example, one company in Louisiana established a 7,100-acre wetland bank, which it anticipated would generate up to US$150 million through the sale of wetland credits. The total documented US market for wetland credits was recently estimated at between US$1.1 billion and US$1.8 billion annually.

Sources: State of Victoria Department of Sustainability and Environment (2006); Hanson et al. (2008); Bayon et al. (2006); Masden et al. (2010)

5.3.2 Voluntary markets for biodiversity and ecosystem services

Voluntary markets for biodiversity and ecosystem services represent an opportunity for businesses to reduce or mitigate their adverse ecological impacts (and potentially to achieve a net positive impact), with the associated benefits of meeting existing or anticipated regulatory requirements, managing environmental risks, and improving corporate environmental and social performance and reputation. Greater rewards may be gained by those companies that exceed public expectations or demonstrate more innovative approaches. Selected examples of voluntary markets for biodiversity and ecosystem services are provided in Box 5.9.

Box 5.9 Selected examples of voluntary markets for ecosystem services

Voluntary Agriculture, Forestry and Other Land Use (AFOLU) carbon credit market: Voluntary carbon transactions reached US$705 million in 2008, doubling in size from the previous year and with credit prices increasing by an average of 20 per cent. Of this total market volume, 11 per cent was accounted for by AFOLU projects, including: afforestation/reforestation, avoided deforestation (REDD), improved forest management, and agriculture soil management. Interest in carbon credits from land-use projects has been driven by both actual and anticipated changes in the regulatory context and the creation of trusted market standards such as the Voluntary Carbon Standard (VCS) and the Climate, Community and Biodiversity Standard (CCBS). These provide assurance for both buyers and sellers, use credible carbon accounting methods and have streamlined registration processes. The methodologies used by voluntary carbon standards do vary, however, with different approaches to quantifying project baselines, leakage and additionality (leakage is when reducing deforestation in one area results in higher deforestation elsewhere; projects are 'additional' when they provide carbon benefits above and beyond what would occur in the 'baseline' scenario). Typical buyers of voluntary, land-based carbon credits include:

- companies that have committed to carbon neutrality;
- companies that purchase carbon credits in order to meet corporate social and environmental responsibility objectives;
- corporate entities that want to anticipate compliance with future regulation;
- traders and brokers speculating on possible future price increases; and
- individuals seeking to offset their personal carbon footprints.

As quality carbon projects are increasingly brought on stream, technical expertise develops and adequate financing is secured, expectations are that the market for AFOLU credits will continue to grow.

The Green Development Initiative (GDI) is a proposed innovative financial mechanism which seeks to engage business in support of the Convention on Biological Diversity (CBD). The GDI aims to mobilize private finance to mitigate biodiversity loss by establishing a standard and accrediting process which will certify the management of geographically defined areas according to CBD objectives and facilitate international payments for GDI-certified biodiversity conservation. It is expected that a voluntary market for GDI-certified areas will enable businesses, consumers and others to finance biodiversity conservation on the ground. For example, if the top 500 companies by market value globally were to devote just one-tenth of 1 per cent of their annual revenues for GDI-certified biodiversity supply, this alone would generate US$24 billion of potential demand.

Mission Markets Inc. is a start-up company predicated on the emergence and growing importance of biodiversity and ecosystem service markets. Mission Markets has created an electronic transactions and communications platform for voluntary environmental and social capital markets, which allows the increasing number of investors focusing on socially responsible investing (SRI) to access a network through which to find, compare and assess new investments, organizations and assets. This provides increased transparency and liquidity, and unifies selected metrics so that investors can compare organizations and assets, confirming the quality and credibility of their transactions.

Malaysia – Malua BioBank is a collaborative effort of the Eco Products Fund LP, a private equity firm jointly managed by New Forests Inc. and Equator Environmental LLC, and the State Government of Sabah, Malaysia, which has granted conservation rights to the Malua BioBank for a period of 50 years. The aim is to raise US$10 million for the rehabilitation of 34,000 hectares of formerly logged forest adjacent to the Danum Valley Conservation Area. The BioBank intends to sell Biodiversity Conservation Certificates, which are each equivalent to 100 square metres of protected and restored rainforest. Certificates are available at US$10 per unit (equivalent to US$1,000 per hectare); revenue from certificate sales is used to fund the running costs of the project and invested in a trust fund for the management of the 50-year licence.

Sources: Hamilton et al. (2008, 2009); Cullen and Durschinger (2008); Gripne (2008); http://gdi.earthmind.net; UNEP/CBD/WG-RI/3/INF/13 (2010); UNEP/CBD/COP/10/INF/15 (2010); www.missionmarkets.com

Taking a lead in voluntary ecosystem service markets can help companies improve their relationships with environmental regulators (BSR 2006). Some regulators may also take lessons from voluntary action, helping to close the gap between regulator expectations and business practices. This can give businesses both influence over and early indications of regulator action.

Ecosystem service markets offer an opportunity to diversify and increase revenue for companies either directly, through project development and the sale of credits in the market, or indirectly, by providing technical support services to facilitate market development and implementation. Although market prospects are difficult to predict and heavily dependent on future regulatory decisions, most markets for ecosystem services are expected to grow rapidly in the coming decades (Mulder et al. 2006). For example, Forest Trends and the Ecosystem Marketplace (2008) forecast that markets for water quality trading under regulatory schemes will exceed US$500 million by 2010. Markets for biodiversity offsets are conservatively estimated to be worth US$1.8 billion to US$2.9 billion today (Madsen et al. 2010), but could grow to US$10 billion by 2020 (Carroll 2008).

At each stage in the operation of ecosystem service markets, there are cost-saving and revenue-generating opportunities for business. Table 5.3 illustrates where these opportunities may lie and the sectors to which they are most relevant.

One of the main commercial opportunities in ecosystem service markets is the design, establishment and management of projects or investments in such projects. Some of the key steps in developing an ecosystem service project are outlined in Figure 5.1. Addressing each of these steps systematically and building flexibility into ecosystem service projects during the early planning stages will help ensure project profitability over the long term.

Existing ecosystem service markets differ significantly in their respective levels of private investment, government regulation and the maturity of supporting institutions. At one end of the spectrum, the US wetland mitigation scheme (initiated in 1983) benefits from clear legal requirements and detailed guidance for wetland offsets in both state and federal government regulations, supported by well-defined liability and enforceable property rights. Market demand for wetland mitigation credits comes from residential, commercial and industrial projects that impinge negatively on wetlands. This in turn has stimulated private investment in the supply of wetland credits and related services (see Box 5.8).

Table 5.3 Business activities supporting ecosystem service markets

Activity	Role	Relevant business sectors
Finance		
Project finance and banking services	Providing investment capital for ecosystem service projects	Investment banks, venture capital, companies seeking to offset negative ecosystem impacts, commercial banks
Fund creation and management	Establishing and managing ecosystem service funds and managing fund investment profiles	Investment fund managers, fund management consultancies
Brokerage	Linking sellers and buyers and facilitating transaction of ecosystem service credits	Brokers and consultancies
Governance		
Monitoring	Collecting and analysing ecosystem service and related information for accountability and transactions statistics	Environmental consultancies, NGOs, research departments
Registry services	Collating and organizing information on an ecosystem service project database	Financial information service companies
Certification	Third-party assurance of performance against certification standards	Environmental consultancies, NGOs, approved certification bodies
Validation and verification	Verifying project plan documents and project performance against market standards	Assurance providers (approved verifiers)
Project Development		
Project developers	Planning, securing finance for and managing the development of ecosystem service projects	Land-owning or land-based companies, construction and infrastructure developers, forestry and agricultural companies, environmental consultancies, private companies
Project technical support	Input of technical expertise and support for the design of ecosystem service projects and services	Environmental consultancies, NGOs, research departments
Market intelligence services	Provision of data on current state of ecosystem service market in exchange for subscription fees	Specialized ecosystem market information providers, news and intelligence agencies, market exchanges, banks
Market strategy support	Interpretation of market information and advice on market strategy	Strategy consultancies, brokerages
Insurance services	Provision of financial compensation for insurable loss and reduction of project risk	Insurance companies
Legal services	Advice on project legal issues, e.g. land tenure, legal protection status, legal status of traded rights	Legal firms

Source: PwC for TEEB

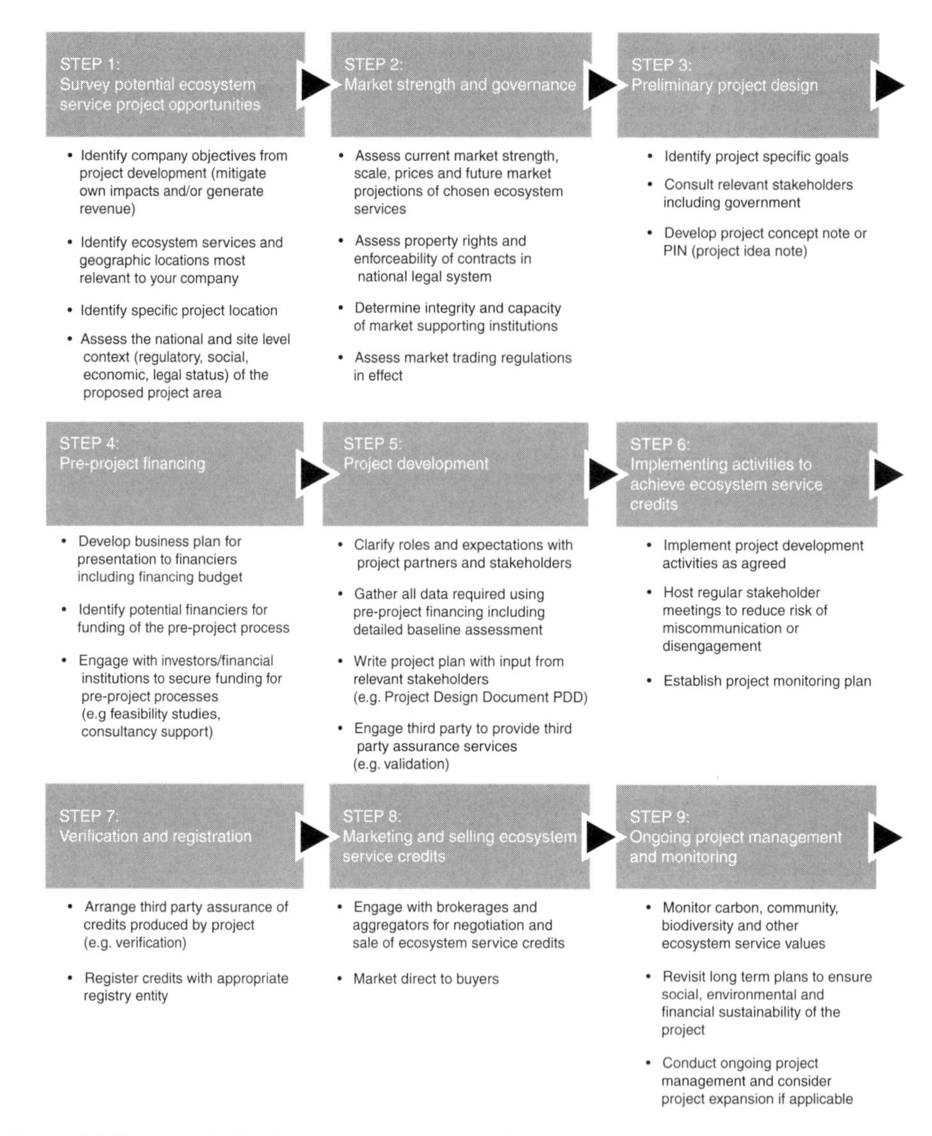

Figure 5.1 *Key steps in developing an ecosystem services project*

Source: PwC for TEEB, incorporating elements from Carter et al. (2009)

At the other end of the spectrum, the potential market for REDD (Reducing Emissions from Deforestation and Forest Degradation) is still awaiting regulatory guidance, development of standardized methodologies, market access agreements, technical support services, and well-defined and enforceable property rights, without which there is likely to be little interest from mainstream investors.

Within the current range of voluntary ecosystem service markets, there are already some significant commercial opportunities. As ecosystem service markets grow, the scale and range of business opportunities is expected to increase substantially. Figure 5.2 identifies some of the key factors that underpin the expansion and strengthening of ecosystem service markets.

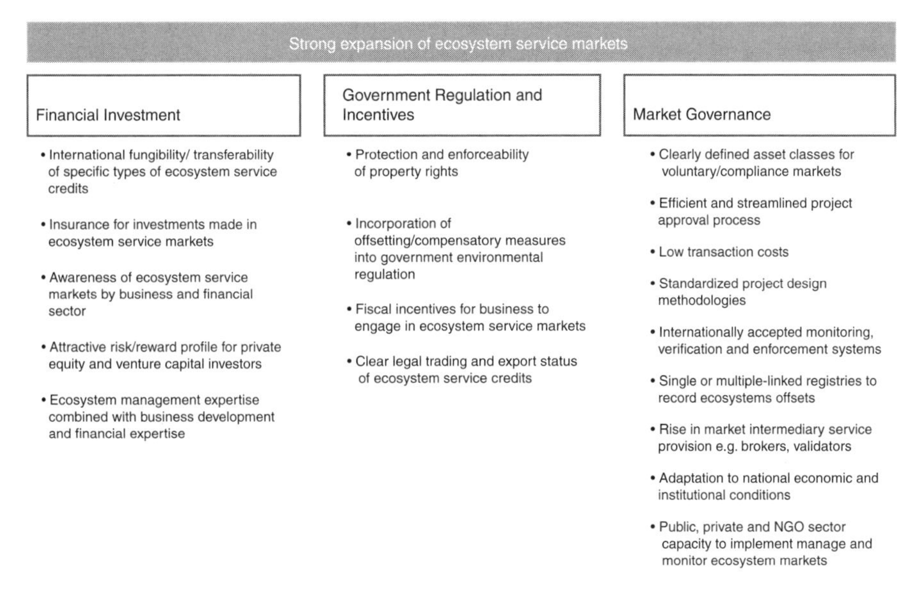

Figure 5.2 *Key factors in the development of markets for ecosystem services*

Source: PwC for TEEB

One of the crucial requirements for developing ecosystem service markets is the establishment and enforcement of well-defined property rights.[3] For an ecosystem project buyer, developer or seller there are a number of related factors to consider, including:

- clear definition of the nature and extent of the property right;
- ability to measure and verify the property right at reasonable cost;
- ability to enforce ownership of the property right at reasonable cost;
- the value of the right and the willingness to purchase the right by other parties;
- ability to transfer ownership of the property right at reasonable cost;
- reliable information on the quality and quantity of ecosystem services provided by the property under current or expected uses; and
- low sovereign risk, meaning that future government decisions are unlikely to significantly reduce the value of property rights.[4]

5.3.3 REDD+ and lessons for other ecosystem service markets

Deforestation and forest degradation is thought to account for around 15 per cent of total anthropomorphic greenhouse gas emissions (Werf et al. 2009). Curbing deforestation and forest degradation appears to offer large, cost-effective and rapid means of mitigating emissions from now until 2030 (Addams et al. 2009; IAP 2009; Copenhagen Accord). A variety of actions intended to achieve this outcome are known as Reducing Emissions from Deforestation and Forest Degradation (REDD). More recently, the scope of REDD has been expanded to 'REDD+', which includes emission reductions from curbing deforestation, plus those from conservation of carbon stocks in standing forests, enhancement of forest carbon stocks, afforestation, reforestation and sustainable management of forests. Landmarks in the development of REDD and REDD+ are indicated in Box 5.10.

Box 5.10 Landmarks in the development of REDD and REDD+

- 2005 Proposal by governments of Costa Rica and Papua New Guinea for inclusion of avoided deforestation in post-Kyoto climate change agreement receives support from 10 countries at COP11 in Montreal
- 2007 United Nations Framework Convention on Climate Change (UNFCCC) parties include reference to REDD in the Bali Action Plan and Bali Road Map for a post-Kyoto climate change framework
- 2009 US$4.6 billion committed to REDD projects across six international funds
- 2009 Recognition of the role and need for a REDD finance mechanism in the COP15 Copenhagen Accord
- 2010 REDD+ mechanism supported by COP16 in Cancun, Mexico
- 2013 REDD+ credits may be allowable compliance units within Phase III of the European Emissions Trading System (EU ETS).

In general, REDD or REDD+ initiatives seek to create or strengthen incentives for the conservation of existing forests through actions that prevent deforestation and/or forest degradation. This may be achieved through various measures, with funding from the carbon market or from governments.

Primary forest is, in general terms, carbon-dense and also biologically diverse, with tropical, temperate and boreal forests containing about two-thirds of the world's terrestrial species. It follows that for intact forest landscapes, one of the most effective ways to conserve biodiversity is by avoiding deforestation. In the case of degraded lands, restoration of forest cover through the establishment of mixed native species along with the avoidance of future deforestation can also yield biodiversity benefits (UNEP 2008).[5]

'Fast-start' funding in the amount of US$4.5 billion has already been committed by the 'REDD+ partnership' nations over the period from 2010 to 2012 to help recipient nations build their REDD strategies, institutional frameworks, public awareness and capacity to deliver REDD. Depending on the impact of this investment and the extent of private sector engagement, REDD+ could develop into the first internationally coordinated, biodiversity-related market of significant size. This market would offer valuable lessons for the development of other ecosystem markets, including how to develop economically efficient, environmentally effective and politically acceptable markets, standards and regulations.

The Eliasch Review (Eliasch 2008) estimated that the funds required to halve emissions from the forest sector by 2030 could be in the range of US$17–33 billion per year, based on estimates from the literature and work commissioned by the review. However, Eliasch further highlighted that the global economic costs of climate change due to deforestation could reach US$1 trillion a year by 2100. In other words, the cost of action is considerably lower than the cost of inaction.

Additionally, the Eliasch Review recommended that REDD+ (along with other forestry mechanisms) should be included in activities to reduce greenhouse gas emissions, as this would significantly lower the overall cost of meeting the emission targets under negotiation in the UNFCCC (Figure 5.3).

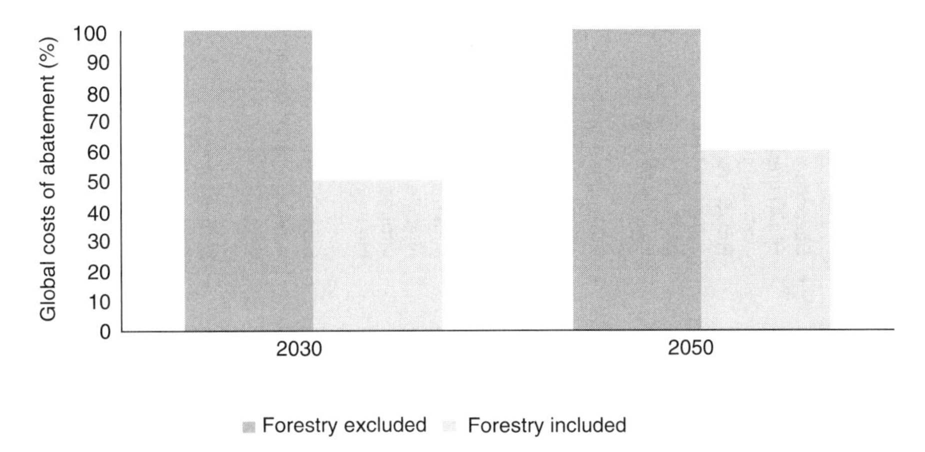

Figure 5.3 *Cost savings including REDD and other forestry mechanisms in climate mitigation*

Source: Eliasch (2008)

Notwithstanding some early voluntary funding commitments, it is difficult to imagine the private sector funding REDD+ on any significant scale without greater clarity about the long-term policy framework and significant strengthening of institutional arrangements in the countries where such activities are expected to take place. Furthermore, concerns about 'leakage' mean that REDD+ is unlikely to be widely accepted unless a coordinated international effort is established. Robust standards and metrics for the environmental and social performance of REDD+ activities will also need to be developed.

Despite the slow pace of climate change negotiations, many governments are keen to demonstrate that REDD+ can work. Several pilot projects have been financed either through voluntary bilateral or multilateral funding, or through the sale of carbon emission reduction credits in the voluntary carbon market. Analysis by PricewaterhouseCoopers for the Conservation Finance Alliance in six countries (Brazil, Indonesia, DRC, Peru, Cambodia and Madagascar) showed that, as of mid-2010, 70 operational REDD projects were publically reported. The analysis also estimated that the total REDD+ funds assigned to date in these six countries is between US$300 million and US$350 million: still a small fraction of the US$4.5 billion in 'Fast-Start' funding pledged for disbursement by 2012 (PricewaterhouseCoopers and Conservation Finance Alliance 2010).

While negotiations continue in the UNFCCC, national and state-level governments continue to forge ahead in the development of carbon markets, with emerging policy in Australia, Japan, South Korea and the US State of California, among others. Early indications suggest that many of these emerging national or sub-national carbon markets will allow the import of credits generated by sub-national REDD+ activities in certain countries for compliance purposes. In other words, REDD+ is expected to offer a significant opportunity to the private sector. Moreover, there are parallel initiatives to develop the supply of other bio-carbon offsets, such as 'blue carbon' in marine ecosystems and 'wet carbon' in wetland ecosystems.

The current immaturity and potential future scale of the REDD+ market presents major opportunities for a variety of businesses. As in other markets, it seems likely that early movers will reap the greatest rewards. Table 5.4 illustrates the range of business opportunities that may arise within different segments of a REDD+ market.

Table 5.4 Business opportunities in REDD+

Finance	Governance	Project development	REDD+ development-related opportunities	Market activity
• Equity investment and project finance – providing upfront financing for establishment of REDD projects, underpinned by carbon revenue streams and grants as well as forest product revenues. • Fund creation and management – creating REDD funds and attracting investment into the fund on a for-profit or not-for-profit basis. Managing the applications, investments, and monitoring of recipient REDD projects.	• Registry services – systems to trade and trace carbon credits generated by projects. • Capacity building for REDD market participants – provision of training and management support for government, civil society organizations and the private sector actors involved in the REDD market. • Validation and verification services – validation and verification of a project's carbon, biodiversity, ecosystem service, community and if relevant, REDD+ performance according to selected standards.	• Project development – planning, securing finance for and managing the development of REDD projects. • Project technical support – technical expertise and support for the design of REDD activities. Creation of technological products to support project design and monitoring (e.g. GIS hardware). • Monitoring – collecting and assessing carbon, biodiversity, ecosystem service, community and other relevant performance data for accountability purposes. • Legal services – advice on contracts (e.g. land ownership). • Insurance services – providing insurance products against risks to reduce risks for REDD activities. • Financial advice – providing advice on project structuring, tax structuring, finance raising and carbon transactions.	• Improving efficiency of forestry operations and energy usage – investment in and development to reduce carbon emissions from forestry operations. • Certified sustainable forest products – production and sale of certified sustainable forest products. • Improved agriculture and livestock technologies – developing or investing in products and services to support efficiency improvements in the livestock and agriculture sectors. • Fuel efficiency and clean energy – developing or investing in products and services to support improvements in fuel and energy efficiency for forestry operations. • Education and training – developing education and training programmes, tools, facilities and support services. • Ecotourism – development of low impact tourism projects in or around REDD project area.	• Secondary trading of REDD credits – purchase and sale of REDD credits on the carbon market for profit. Providing liquidity to a market. • Brokerage – linking sellers and buyers and facilitating REDD credit transactions. • Project aggregation – generating portfolios of multiple REDD projects through purchasing in the primary market. • Market intelligence services – provision of data on REDD market. • Market technical support – interpretation of market information and giving advice on marketing/sales strategy.

Source: PwC for TEEB, building on Forum for the Future (2009)

While REDD+ could deliver both climate and biodiversity benefits, it is also clear that some industrial sectors may be negatively affected if deforestation is significantly reduced. This will be particularly acute for businesses that rely on access to rural land, such as agriculture or forest, paper & packaging (FPP). These industries could see increased competition for access to natural forests and other lands from carbon developers seeking to establish REDD+ projects. This could result in higher land prices, limit the area available for timber harvesting and increase operational costs. Such impacts may be offset to some extent by companies through active participation in the REDD+ market. For example, FPP companies may afforest or reforest land and implement sustainable forest management systems for the purposes of forest carbon stock enhancement and generate revenue through the sale of carbon offset credits, alongside pulp and paper products.

Regulatory uncertainties aside, the potential benefits of REDD+ in both the short and long term have led some businesses to take the plunge. Box 5.11 describes how a large international hotel chain is seeking to offset its greenhouse gas emissions by funding a REDD scheme in the Brazilian Amazon.

Box 5.11 Marriott invests in REDD: The Juma Sustainable Development Reserve

In Brazil, the Marriott international hotel group has dedicated US$2 million to a fund administered by the Amazonas Sustainable Foundation, which protects 590,000ha of threatened rainforest in the Juma Sustainable Development Reserve. The hotel group is investing in this project partly in order to offset its carbon footprint of about 3 million metric tons of CO_2 per year. The initiative has helped the company win numerous sustainability awards, including the World Travel and Tourism Council's 2009 'Tourism for Tomorrow Award for Sustainability'.

The Juma project provides incentives for the local population to protect the forest by distributing payments through the 'Bolsa Floresta' programme. For example, the 'Bolsa Floresta Family' programme provides a value card for each local family, which can be credited with 50 Reais (around US$25) each month dependent on forest protection performance. Investments in the project are also used to fund research and conservation work in the reserve, as well as supporting local economic initiatives. If all goes to plan, the project will avoid deforestation of 330,000 ha of natural rainforest over the period to 2050, generating carbon credits of around 189 million tons. The reputational boost gained by the hotel group's investment in the Juma project has been significant and is expected to increase their appeal to the sustainable tourism market.

Sources: www.forestcarbonportal.com; www.marriott.co.uk; Forum for the Future (2009); PwC for TEEB

5.4 Tools to support markets for biodiversity and ecosystem services

Biodiversity business opportunities are supported by an increasingly robust set of market-based tools. These enable companies to adopt biodiversity-responsible practices and to develop and market biodiversity-based goods and services. Such market-based tools complement the many policy instruments for biodiversity reviewed in the TEEB volume for national and international policy makers (TEEB in Policy Making 2011). This section highlights market-based tools for biodiversity business, focusing on those related to responsible investment, production and reporting.

For the investment community, a key mechanism for integrating biodiversity in decision making is Performance Standard 6 on Biodiversity Conservation and Sustainable Management of Living Natural Resources, developed and recently updated by the International Finance Corporation (IFC). This standard, together with all of the IFC's environmental and social standards, not only guides the investments of the IFC – the private sector arm of the World Bank – but also influences the investment practices of more than 60 multinational banks that have adopted the Equator Principles (TEEB in Policy Making 2011). The Equator Principles call for adherence to IFC Performance Standards with respect to project financing in excess of US$10 million in emerging markets. In the recent updating of the IFC Performance Standards the second iteration of Performance Standard 6, in particular, aims to provide an even more robust framework for addressing biodiversity issues in investment decisions.[6]

5.4.1 Certification schemes for biodiversity and ecosystem services

Growing concern about environmental and social issues has stimulated demand for products and production practices that conserve biodiversity and use biological resources more sustainably and equitably. To meet this demand, a range of voluntary standards and certification schemes have been developed, which aim to verify the environmental and social aspects of products and production processes. Although few existing voluntary standards and certification schemes focus specifically on biodiversity, most address some aspects of the biodiversity challenge.

For example, in the coffee sector, there are schemes that emphasize landscape or ecosystem protection, such as Rainforest Alliance certification, while others promote environment-friendly farming practices (e.g. organic agriculture), and still others emphasize social equity in the use of biological resources (e.g. Fairtrade).

Taken as a whole, such certification schemes can contribute to biodiversity-responsible business practices. A few schemes, however, are particularly noteworthy in terms of their contributions to biodiversity conservation at the landscape and ecosystem level:

- Rainforest Alliance certification is a comprehensive process that promotes and guarantees improvements in agriculture and forestry. The scheme aims to ensure that the goods and services certified were produced 'in compliance with strict guidelines for protecting the environment, wildlife, workers and local communities' (www.rainforest-alliance.org).[7]
- The Marine Aquarium Council (MAC) brings together fisheries and organizations that collect, produce and handle marine aquarium organisms around the world, committing its members to work towards compliance with a shared set of standards. The MAC vision is to certify the entire supply chain, starting with sustainable and responsible management of the marine site from which fish and marine organisms are harvested. The MAC Ecosystem Fishery Management Standard includes a commitment: 'to verify that the collection area is managed

according to principles of ecosystem management in order to ensure ecosystem integrity and the sustainable use of the marine aquarium fishery'.

- The LIFE Institute in Brazil has developed 'LIFE Certification', which qualifies and recognizes public and private organizations that promote biodiversity conservation and sustainable development initiatives, with the aim of promoting business support for ensuring the protection of ecosystem integrity.

These examples are indicative of the growing number of biodiversity-related voluntary standards and certification schemes. Others include the Climate, Community and Biodiversity Alliance, the International Federation of Organic Agriculture Movements, the FairWild Foundation, the Forest Stewardship Council, the Marine Stewardship Council, the Roundtable on Sustainable Palm Oil and the Union for Ethical BioTrade. Each of these voluntary certification schemes provides structures, standards and processes that help companies validate their biodiversity responsibility to consumers and other stakeholders in the market. In this way, such schemes help companies to penetrate more discerning markets and become more attractive to the growing number of concerned consumers and investors.

Many social and environmental certification schemes are increasingly consistent in their structure and processes, in part due to the sharing of experience and lessons learned (see Box 5.12). Nevertheless, while all certification schemes emphasize the importance of regular audits and independent evaluations of the environmental and social performance of their subscribers, the schemes themselves are only rarely subjected to such independent assessments. One recent review of the impacts of

Box 5.12 ISEAL: The association for social and environmental standards

ISEAL (International Social and Environmental Accreditation and Labelling) Alliance is the global association for social and environmental standards and has become an umbrella organization for a growing number of standards and certification schemes. Members include Fairtrade Labelling Organizations International (FLO), Forest Stewardship Council (FSC), International Federation of Organic Agriculture Movements (IFOAM), International Organic Accreditation Service (IOAS), Marine Aquarium Council (MAC), Marine Stewardship Council (MSC), Rainforest Alliance and Social Accountability International (SAI). One of the key roles of ISEAL is to harmonize the way that social and environmental standards for goods and services are administered, verified and assessed.

To this end, ISEAL developed a 'Verification Code of Good Practice for Assessing the Impacts of Standards Systems'. This defines good operating practices in terms of accreditation, certification and auditing to social and environmental standards, and creates a requirement for all standards systems to measure and demonstrate their contributions to social and environmental impacts using consistent methodologies. A key aim of the code is to strike a balance between rigorous certification that meets the expectations of consumers while also ensuring that smaller businesses can afford to enter certification programmes and perceive them as market enablers.

Source: www.isealalliance.org

'sustainable' certification found only modest empirical support for claims that such certifications deliver significant social and environmental benefits (Blackman and Rivera 2010). Out of 37 studies considered, the authors identified 14 (focusing on bananas, coffee and tourism) that were considered likely to generate 'credible' results based on sound ex-post analysis and plausible counter-factual scenarios. Among these 14 studies, six appeared to indicate socio-economic benefits from 'sustainable' certification, but only limited evidence of environmental benefits.

Another study of the impact of certification focused on sustainability initiatives in the coffee sector (Giovannucci and Potts 2008). The authors present a methodology to assist farmers and others in assessing and predicting the social, economic and environmental outcomes they may expect from implementing various sustainability initiatives (including Fairtrade, organic, UTZ-certified and Rainforest Alliance). In the process of developing and testing the methodology on 50 farms in five countries, the authors generate 'preliminary' observations, which suggest that:

- farm performance along social, economic and environmental indicators is highly variable, mainly reflecting local conditions;
- certified farms appeared to be better off than their conventional counterparts, at least in terms of net farm income;
- certification had not delivered significant environmental benefits, at least over the first two years of certification; and
- certified farms appeared to have better occupational health and safety, employee relations and labour rights performance.

Perhaps the main lesson to be drawn from evaluations of 'sustainable' certification is the need for more systematic analysis of the social and environmental effectiveness of such schemes.

5.4.2 Assessment and reporting for biodiversity and ecosystem services

As companies begin to address biodiversity more strategically in their planning and decision making, new tools for assessing and reporting on biodiversity are needed, as discussed in Chapter 3. Below we highlight three examples that show how such tools can help companies realize new biodiversity business opportunities:

- The Corporate Ecosystem Services Review (ESR) provides a wealth of tools and training materials to assist companies in addressing the biodiversity aspects of their operations. For example, Mondi 'used the ESR to develop several new strategies for dealing with the ecosystem service challenges to their FSC certified South African plantations', while Syngenta used the ESR to 'identify a number of possible opportunities to help farmers either reduce their impacts on ecosystems or adapt to ecosystem change'.
- The Global Reporting Initiative (GRI) Biodiversity Reporting Resource was released in 2007 to assist companies in reporting their biodiversity performance. As the tool documentation explains: 'reporting offers organizations an opportunity to explain their relationship with biodiversity. In what way does the organization respond to negative impacts on biodiversity deriving from its activities? What are positive impacts?'
- The Natural Value Initiative (NVI) helps the finance sector evaluate and assess how investee companies – particularly in the food, beverage and tobacco sectors – are managing biodiversity opportunities and risks (www.naturalvalueinitiative.org). It uses the Ecosystem Services Benchmark (ESB), which is somewhat akin to the ESR noted above. The ESB has

been tested on over 30 companies that are particularly dependent on biodiversity and can therefore have significant impacts on mitigating biodiversity loss.

5.4.3 Voluntary incentives for biodiversity business

As discussed in the TEEB volume for national and international policy makers, market incentives for biodiversity and ecosystem services can be strengthened through appropriate public policies and regulations (TEEB in Policy Making 2011). However, in addition to these policy tools, there are many voluntary incentives which can encourage and enable biodiversity business. These include:

- awareness-raising initiatives, which can influence the perceptions of investors, managers, employees or consumers about biodiversity;
- voluntary biodiversity offsets, which are conservation measures that can compensate for the residual, unavoidable harm to biodiversity caused by development projects;
- Biotrade arrangements, which promote the collection, production, transformation and commercialization of goods and services derived from biodiversity under criteria of environmental, social and economic sustainability;
- bioprospecting arrangements, and access and benefit-sharing agreements, which can fund biological research and discovery partnerships between pharmaceutical companies and countries with high levels of biodiversity;
- stewardship payments, which reward resource managers for delivering biodiversity conservation and providing ecosystem services that benefit the public;
- conservation auctions, which can be used to deliver stewardship payments;
- conservation covenants, as part of voluntary land transactions, to embed biodiversity conservation in the title deed;
- payments for watershed protection, ranging from payments by private water users to environmental agencies and NGOs, to direct payments by central governments to private landowners; and
- public–private partnerships, which explore business opportunities for safeguarding biodiversity and can link commercial debt finance with public subsidies to produce social, commercial and biodiversity benefits.

Using these and other approaches, leading companies, as well as public and non-profit organizations, are developing new ways to stimulate business engagement in the conservation and sustainable use of biodiversity and ecosystem services.

5.4.4 Further considerations for institutional investors

Most institutional investors have little understanding or see little value in biodiversity and ecosystem services. This could change as biodiversity and ecosystem services are increasingly valued by markets, causing capital inflows to entrepreneurs and companies that know how to seize new biodiversity business opportunities, as well as disinvestment from companies that are perceived to neglect biodiversity and ecosystem services.

F&C, an asset manager, identified the following policy characteristics relevant to the development of biodiversity and ecosystem service trading and related market opportunities (F&C Investments 2010):

- In order to reduce investment risk, investors need a level of certainty about the fundamental nature of a new market – this is likely to require government intervention and regulation.
- Investors are wary of subsidies that may be subject to swift changes due to political pressure – such as the EU subsidies for bio-fuels.
- Landowners and those who grant permits for development could create rapid changes in the effectiveness of biodiversity management by linking development permit awards to good environmental management.
- All ecosystem services cannot be traded with ease, and so it will be necessary to focus on a restricted number that can familiarize investors and markets with the concepts and allow an understanding of the risks and opportunities. The importance of building knowledge and trust among investors cannot be overstated; excessively complex or rushed schemes are likely to fail.
- Changes should not unfairly disadvantage companies based in particular countries – for example, increasing the costs of access to land for a mining company should apply equally to all such companies, whatever their domicile.
- Data provision is crucial to good investment decisions, and also to good public policy decisions.
- Cost-effective metrics and valuation methodologies are needed – for example, to value the ecosystem services delivered by forests in monetary terms, or to assess ecosystem service debits and credits in bio-physical terms.

If public policy reforms intended to encourage biodiversity business are to attract significant support from institutional investors, concerns such as these will need to be addressed comprehensively.

5.4.5 Public policy to support biodiversity business

The TEEB volume for policy makers includes a detailed review of various policy interventions that can and, in some cases, have been implemented at international and national levels in an effort to reduce the loss of biodiversity (TEEB in Policy Making 2011). Some of these policies would create an enabling framework to scale up new biodiversity and ecosystem service business opportunities, notably the policies discussed in 'Rewarding Benefits through Payments and Markets' (TEEB in Policy Making 2011: chapter 5), such as:

- Public and private payments for ecosystem services (PES), where effective targeting, monitoring and governance – and measures to ensure additionality (beyond 'business as usual') and minimize leakage (displacing damage elsewhere) can ensure fairness and value for money.
- International PES opportunities, which include the financial mechanism proposed with the post-2012 regime under the UNFCCC, REDD+, a proposed Green Development Initiative (which aims to support the CBD in its work on innovative financial mechanisms) and other emerging initiatives.
- Tax systems and intergovernmental fiscal transfers that can provide positive incentives for private actors and public agencies to make investment choices favouring long-term stewardship over short-term development options.
- Historically, host countries have benefited little from the development and commercialization of products based on traditional knowledge and genetic

resources. Fairer and more efficient contractual mechanisms can recognize value, establish rights for local people and facilitate discovery and application across many sectors.
- Policy makers can support certification and labelling schemes for a wider range of products and services by backing robust standards and verification systems that cover biodiversity conservation, and providing advice and capacity services for business.

Such policy reforms, if widely adopted, would reinforce the voluntary tools and incentives for biodiversity business outlined above. In order to maximize synergies between private initiatives and public policies, however, it is essential for business to be more actively involved in the development and review of biodiversity policies, at both international and national levels.

5.5 Challenges to building biodiversity business

While there is solid evidence and growing practical experience of biodiversity and ecosystem related opportunities for business, there remain significant challenges to building biodiversity businesses. Some of these challenges are conceptual, others are technical and still others are political or even ideological.

At a conceptual level, the very word 'biodiversity' is often difficult for business people to understand. What exactly is biodiversity? What are we supposed to do about it? How does biodiversity relate to the concept of 'ecosystem services' and what exactly is business supposed to do about these services? These are not trivial questions, if business is to be expected to produce, market, sell and manage goods and services in ways that are more compatible with biodiversity conservation and help sustain or enhance ecosystem services.

An important reference for understanding biodiversity and what business can do about it is the Convention on Biological Diversity (CBD). As the major international agreement on biodiversity, the CBD offers clear guidance for business and others. The Preamble of the Convention and Articles 1 and 2 identify the core components of biodiversity as:

- ecological complexes (such as landscapes),
- ecosystems,
- species, and
- biological resources (including genetic resources).

It is also clear from the CBD what society – including business – is supposed to do:

- conserve biodiversity,
- use biological resources sustainably,
- ensure such use is equitable (particularly with respect to genetic resources), and
- promote economic to social development to eradicate poverty.

The technical challenges facing biodiversity business are significant and include the challenge of assessing, measuring, monitoring and reporting on the biodiversity performance and dependence of business activities. As discussed in Chapter 3, one reason why businesses historically have not addressed biodiversity in a systematic way is that it can be difficult to measure the links between business, biodiversity and ecosystem services.

Attempts to express business impacts and dependence on biodiversity in simple measurable units, much as we measure greenhouse gas emissions in terms of carbon dioxide equivalent units, continue to prove terribly challenging. Perhaps other metrics (or indeed other types of business action) are required. For example, the Green Development Initiative (GDI) is exploring how to certify landscapes managed for biodiversity in simple per hectare terms, rather than attempting to measure the 'amount' of biodiversity that is conserved. In short, biodiversity cannot be treated as a commodity like carbon but requires a different approach. A better analogy may be markets for art, property or land, which can and do attract significant investment, but are not normally treated as interchangeable commodities (http://gdi.earthmind.net).

Political challenges also create significant barriers to the development of biodiversity businesses and markets. Most biodiversity assets – including mountains, forests, grasslands, wetlands, lakes and rivers, coastal areas and exclusive economic zones in the oceans – are currently owned and managed by governments. It is very difficult, if not impossible, for sustainable biodiversity businesses to develop around assets over which they have no or only limited use rights. Moreover, while political leaders would like the private sector to play a greater role in conserving biodiversity and sustaining ecosystem services, the political will to 'privatise the management of nature' is simply not evident. Indeed, politicians in some countries, supported by a mix of NGOs, academics and community-based organizations, remain vehemently opposed to what they see as the 'commodification of nature'.

The issue boils down to a question of trust. Can we trust business to conserve biodiversity and sustain ecosystem services? Do we trust business to deliver public goods in their quest for private profits? Do we believe that business leaders will do what our political leaders have so far failed to achieve, and halt biodiversity loss?

The short answer for many people is no. This does not necessarily reflect a deep-seated opposition to markets for biodiversity or to business making profits from producing goods and services in a responsible way. Rather, many people do not believe that the private sector can deliver real and measurable biodiversity results. This is because the business case for biodiversity conservation is still not obvious to many people. Further, we may suspect that business action on biodiversity is little more than 'greenwashing'. No doubt in some cases this is accurate. Nevertheless, with biodiversity continuing to decline at an alarming rate, and with governments apparently unable or unwilling to assume responsibility for biodiversity conservation, it seems increasingly necessary to encourage business to play a much greater role in the collective effort to protect, conserve and restore the natural resources upon which we all depend.

The real challenge to building biodiversity business is us. Can we, as citizens, as consumers, and also as investors, managers and workers, call on business to take a leadership role in conserving biodiversity and sustaining ecosystem services? Can we support efforts ranging from biodiversity policy reforms and investment plans, to voluntary biodiversity standards and certification schemes, to third-party monitoring and reporting of biodiversity performance, which will enable and encourage business not only to engage more fully in the biodiversity agenda, but ultimately to become biodiversity positive?

5.6 What is to be done?

There are new approaches for integrating biodiversity into existing businesses; new biodiversity business models are emerging; and new markets for biodiversity or ecosystem services are being developed. This chapter has sought to demonstrate that biodiversity and ecosystem services are more than just a risk to business; they are increasingly the basis of profitable and sustainable business opportunities.

This chapter proposes that a business-like approach to biodiversity and ecosystem services can deliver positive results for both biodiversity and the bottom line. As suggested in Table 5.5, biodiversity and ecosystem services are likely to provide increasingly lucrative business opportunities in the years ahead.

From a business perspective, the opportunities related to biodiversity and ecosystem services are increasingly compelling. These opportunities are most evident where business profits depend directly on the quality and quantity of ecosystem services – in ecotourism ventures, for example – but also for businesses that rely

Table 5.5 Current and expected value of biodiversity-related markets and finance

Market opportunities and other biodiversity finance	Actual 2010 (US$ millions)	Estimated 2013 (US$ millions)	Estimated 2020 (US$ millions)
Nature-based tourism and recreation	80,000	90,000	200,000
Certified agricultural products (e.g. organic, conservation grade)	56,000	98,000	261,000
Certified forest products (FSC only)	30,000	45,000	228,000
Payments for water-related ecosystem services (government)	9,000	12,000	20,000
Mandatory biodiversity offsets (e.g. US mitigation banking)	3,000	4,000	5,000–8,000
Water quantity permit trading	3,000	3,750	4,800
Other payments for ecosystem services (government-supported)	2,200	2,300	2,900
Publicly financed REDD+	1,000	<7,000	500–15,000
Forest carbon (voluntary markets)	54.7	50	<5,000
Forest carbon (compliance markets)	26.9	<5,000	<5,000
Payments for genetic resources	35	35	100
Voluntary biodiversity offsets	24–27	30	70
Payments for watershed management (voluntary)	5	10	50
Traditional conservation finance (government spending, overseas aid and philanthropy)	29,000–35,000	N/A	N/A

Sources: Adapted from: http://moderncms.ecosystemmarketplace.com/repository/moderncms_documents/PES_MATRIX_06-16-08_oritented.1.pdf, updated through personal communications with Ecosystems Marketplace (2011)

heavily on biological resources such as wood, water, fibre, fish and wild genetic material. As noted above, biodiversity business opportunities can be found in a range of sectors, including agriculture, banking, consulting, construction, cosmetics, education, energy, fashion, fisheries, forestry, housing, mining, pharmaceuticals, retail, publishing, tourism and transportation.

So what is to be done? Why do we not see more private investment in biodiversity and ecosystem services? What is holding back the growth of biodiversity businesses across a range of sectors? If biodiversity and ecosystem services are as important to life on earth as scientists tell us, why do we not see more rapid expansion of markets for ecosystem services?

Certainly we need more and better biodiversity management tools for business. Perhaps more importantly, however, we need to develop more confidence in the potential of business to conserve biodiversity and supply ecosystem services. In practice, this implies not just assigning more responsibility to business for conserving biodiversity but also, along with this increased responsibility, entrusting business with access to biodiversity assets and more influence on how these assets are managed.

In short, if we want to harness the full power of business to help stop the loss of biodiversity and restore degraded landscapes and ecosystems, we will need to rethink the way societies manage natural capital and biodiversity assets.

Acknowledgements

Stuart Anstee (Rio Tinto), Andrea Athanas (IUCN), Bruce Aylward (Ecosystem Economics), Ricardo Bayon (EKO Asset Management Partners), Maria Ana Borges (IUCN), Roberto Bossi (ENI) David Brand (New Forests), Jim Cannon (Sustainable Fisheries Partnership), Nathaniel Carroll (Ecosystem Marketplace), Catherine Cassagne (IFC and Sustainability Advisory Services), Sagarika Chatterjee (F&C Investments), Ian Dickie (Eftec), Steinar Eldoy (StatoilHydro), Eduardo Escobedo (UNCTAD), Jan Fehse (EcoSecurities), Sean Gilbert (Global Reporting Initiative), Marcus Gilleard (Earthwatch Institute Europe), Annelisa Grigg (Global Balance), Frank Hicks (Biological Capital), Ard Hordijk (Nyenrode Business University), Joël Houdet (Orée), Mikkel Kallesoe (WBCSD), Sachin Kapila (Shell), Becca Madsen (Ecosystem Marketplace), Nadine McCormick (IUCN), Andrew Mitchell (Global Canopy Programme), Jennifer Morris (Conservation International), Carsten Neβöver (UFZ), Bart Nollen (Nollen Group), Ashim Paun (Cambridge University), Paola Pedroni (ENI), Danièle Perrot-Maître (UNEP), Wendy Proctor (CSIRO), Mohammad Rafiq (Rainforest Alliance), Conrad Savy (Conservation International), Paul Sheldon (Natural Capitalism Solutions), Daniel Skambracks (KfW Bankengruppe), Dale Squires (UC San Diego), Alexandra Vakrou (European Commission), Jon Williams (PricewaterhouseCoopers).

Notes

1 IFOAM (International Federation of Organic Agriculture Movements) explains: 'Organic agriculture is a production system that sustains the health of soils, ecosystems and people. It relies on ecological processes, biodiversity and cycles adapted to local conditions, rather than the use of inputs with adverse effects. Organic agriculture combines tradition, innovation and science to benefit the shared environment and promote fair relationships and a good quality of life for all involved.' URL: www.ifoam.org/press/press/2008/20080522_Press_Release_Organic_Agriculture_for_Biodiversity.php

2 A recent survey by UEBT (Union for Ethical BioTrade) indicated significant consumer interest in biodiversity in this sector. URL: http://uebt.ch/conferences/dl/UEBT_BIODIVERSITY_BAROMETER_web280410.pdf

3 Property rights can be defined as a 'claim to a benefit (or income) stream that the State will agree to protect through the assignment of duty to others who may covet, or somehow interfere with, the benefit stream' (Bromley 1991). For business, the protection of this claim to environmental service (or

income) streams is fundamental to engaging in ecosystem service markets (Adger and Luttrell 2000; Bayon 2004).

4 These factors may differ according to the particular ecosystem service or country in which an investment is being made.

5 In principle, if REDD+ activities taken place in areas of both high carbon stocks and high biodiversity values, then they will simultaneously help meet the emission reduction objectives of the UNFCCC as well as the conservation and sustainable use objectives of the Convention on Biological Diversity (UNEP 2008).

6 The April 2010 draft states: 'Performance Standard 6 recognizes that protecting and conserving biodiversity, the maintenance of ecosystem services, and the sustainable management of natural resources are fundamental to sustainable development. This Performance Standard reflects the objectives of the Convention on Biological Diversity to conserve biological diversity and promote the use of renewable natural resources in a sustainable manner. This Performance Standard addresses how clients can avoid, reduce, restore, and offset impacts on biodiversity arising from their operations as well as sustainably manage renewable natural resources and ecosystem services' (www.ifc.org/ifcext/policyreview.nsf/Content/Performance-Standard6).

7 Two key criteria with respect to ecosystems in the Rainforest Alliance standard for sustainable agriculture are (www.rainforest-alliance.org/agriculture/documents/sust_ag_standard.pdf):

- 2.1 All existing natural ecosystems, both aquatic and terrestrial, must be identified, protected and restored through a conservation program. The program must include the restoration of natural ecosystems or the reforestation of areas within the farm that are unsuitable for agriculture.
- 2.2 The farm must maintain the integrity of aquatic or terrestrial ecosystems inside and outside of the farm, and must not permit their destruction or alteration.

References

Addams, L., Boccaletti, G., Kerlin, M. and Stuchtey, M. (2009) *Charting Our Water Future: Economic Frameworks to Inform Decision Making*, 2030 Water Resources Group, McKinsey & Company. URL: www.mckinsey.com/clientservice/water/charting_our_water_future.aspx

Adger, W.N. and Luttrell, C. (2000) 'Property rights and the utilization of wetlands', *Ecological Economics*, 35(1): 75–89.

Barber, T. and Krivoshlykova, M. (2006) *Global Market Assessment for Handicrafts*, Volume 1 (final draft), USAID, Washington, DC.

Bayon, R. (2004) 'Making environmental markets work: Lessons from early experience with sulphur, carbon, wetlands and other related markets', presented at Katoomba Group meeting, Lucarno, Switzerland.

Bayon, R., Carroll, N., Hawn, A., Kenny, A., Walker, C., Bruggeman, D., et al. (2006) *Banking on Conservation: Species and Wetland Mitigation Banking*, The Katoomba Group's Ecosystem Marketplace. URL: http://moderncms.ecosystemmarketplace.com/repository/moderncms_documents/market_insights_banking_on_mitigation.1.pdf

Bernick, L. and Guth, J. (2010) 'Retail: Stocking the shelves with green', *GreenBiz Reports*, Five Winds International (March).

BioFach (2009a) 'Weleda on the road to success', *BioFach – Vivaness Newsletter*, no. 203 (7 August).

BioFach (2009b) 'Natural cosmetic growing against the trend', *BioFach – Vivaness Newsletter*, no. 204 (21 September).

Bishop, J., Kapila, S., Hicks, F., Mitchell, P. and Vorhies, F. (2008) *Building Biodiversity Business*, Shell International Ltd and IUCN, London, UK, and Gland, Switzerland (March). URL: http://data.iucn.org/dbtw-wpd/edocs/2008-002.pdf

Blackman, A. and Rivera, J. (2010) *The Evidence Base for Environmental and Socioeconomic Impacts of 'Sustainable' Certification*, Environment for Development Discussion Paper Series, EfD DP 10-10 (March).

Bromley, D.W. (1991) *Environment and Economy: Property Rights and Public Policy*, Basil Blackwell, Oxford.

BSR (2006) *Environmental Markets: Opportunities and Risks for Business*. URL: www.bsr.org/reports/BSR_Environmental-Markets.pdf

Carbon Disclosure Project (CDP) (2010) *Becoming a Signatory or Founding Signatory Member*. URL: www.cdproject.net/SiteCollectionDocuments/CDP_WD_Signatory_Brochure_2010.pdf

Carroll, N. (2008) 'Compliant biodiversity offsets', in *Payments for Ecosystems Services: Market Profiles, Forest Trends and the Ecosystem Marketplace*. URL: http://ecosystemmarketplace.com/documents/acrobat/PES_Matrix_Profiles_PROFOR.pdf

Carter, I.J., Stevens, T., Clements, T., Hatchwell, M., Krueger, L., Victurine, R., et al. (2009) *WCS REDD Project Development Guide*, USAID. URL: www.translinks.org/Docustore/tabid/409/language/en-GB/Default.aspx?Command=Core_Download&EntryId=3646

CBI (2004) European buyers' requirements: Benchmarking the tourism industry (September).

CBI (2008) The Outerwear market in the EU, Prepared by Fashion Research & Trends (September).

CBI (2009a) Long Haul Tourism: The EU market for adventure travel (March). URL: www.cbi.eu/marketinfo/cbi/docs/long_haul_tourism_the_eu_market_for_wildlife_travel

CBI (2009b) Luggage and (leather) accessories – CBI Market Survey: The EU Market for wallets and purses (April).

CBI (2009c) European buyer requirements: Tourism (June). URL: www.cbi.eu/marketinfo/cbi/docs/european_buyer_requirements_tourism

Clean Edge (2009) *Clean Energy Trends 2009* (March). URL: www.cleanedge.com/reports/pdf/Trends2009.pdf

Connor, M. (2010) 'Survey: US Consumers Willing to Pay for Corporate Responsibility', *Business Ethics, the Magazine of Corporate Responsibility* (29 March). URL: http://businessethics.com/2010/03/29/1146-survey-u-s-consumers-willingto-pay-for-corporate-responsibility/

Cullen, M.A. and Durschinger, L.L. (2008) Emerging market for land-use carbon credits, ITTO Tropical Forest Update 18/3. URL: www.itto.int/direct/topics/topics_pdf_download/topics_id=1881&no=0

Ecosystem Services Project (2008) *The Markets for Ecosystem Services Project: Factsheet*. URL: www.ecosystemservicesproject.org/html/publications/docs/facts/Markets_Flyer2_web.pdf

Eliasch, J. (2008) *Climate Change: Financing Global Forests: The Eliasch Review*. Earthscan, London.

Evison, W. and Knight, C. (2010) *Biodiversity and Business Risk: A Global Risks Network Briefing*, World Economic Forum, Geneva. URL: www.weforum.org/pdf/globalrisk/Biodiversityandbusinessrisk.pdf

F&C Investments (2010) *Response by F&C Investments to consultation on The Economics of Ecosystems & Biodiversity (TEEB)*, URL: www.fandc.com/FN_FileLibrary/file/FC_submission_to_The_Economics_of_Ecosystems_Biodiversity_TEEB_study.doc

Fair Trade in Tourism South Africa (FTTSA) (2009) personal communication.

Fairtrade Labelling Organization (2010) 'Facts and figures' web page. URL: www.fairtrade.net/facts_and_figures.html

FiBL and IFOAM (2009) *The World of Organic Agriculture, Statistics and Emerging Trends 2009*, Frick, Bonn and Geneva.

Fischer, C. (2003) *Combining Rate-Based and Cap-and-Trade Emissions Policies*, Discussion Paper, Resources for the Future. URL: www.rff.org/Documents/RFF-DP-03-32.pdf

Forest Trends and the Ecosystem Marketplace (2008) *Payments for Ecosystem Services: Market Profiles*. URL: http://ecosystemmarketplace.com/documents/acrobat/PES_Matrix_Profiles_PROFOR.pdf

Forum for the Future (2009) Forest Investment Review. URL: www.forumforthefuture.org/projects/forest-investmentreview

Giovannucci, D. and Potts, J., with Killian, B., Wunderlich, C., Soto, G., Schuller, S., Pinard, F., Schroeder, K. and Vagneron, I. (2008) *Seeking Sustainability: COSA Preliminary Analysis of Sustainability Initiatives in the Coffee Sector*, Committee on Sustainability Assessment, Winnipeg.

Global Industry Analysts, Inc. (2008) *Agricultural Biotechnology: A Global Strategic Business Report*, Global Industry Analysts, Inc., San Jose, CA.

Global Reporting Initiative (GRI) (2007) *Biodiversity: a GRI Reporting Resource*. URL: www.globalreporting.org/NR/rdonlyres/07301B96-DCF0-48D3-8F85-8B638C045D6B/0/BiodiversityResourceDocument.pdf (last accessed 28 June 2010).

Gripne, S. (2008) 'Markets for biodiversity: Delivering returns from emerging environmental markets', *PERC Reports*, vol. 26, no. 4 (December).

Hamilton, K., Sjardin, M., Marcello, T. and Xu, G. (2008) *Forging a Frontier: State of the Voluntary Carbon Markets 2008*, Ecosystem Marketplace and New Carbon Finance. URL: www.ecosystemmarketplace.com/documents/cms_documents/2008_StateofVoluntary CarbonMarket2.pdf

Hamilton, K., Sjardin, M., Shapiro, A. and Marcello, T. (2009) *Fortifying the Foundation: State of the Voluntary Carbon Markets 2009*, The Katoomba Group and New Carbon Finance.

Hanson, C., Ranganathan, J., Iceland, C. and Finisdore, J. (2008) *The Corporate Ecosystem Services Review: Guidelines for Identifying Business Risks and Opportunities Arising from Ecosystem Change*, WRI, WBCSD and Meridian Institute, Washington, DC. URL: http://pdf.wri.org/corporate_ecosystem_services_review.pdf

HSBC (2009) Climate Change Index annual review (September).

IAP (2009) 'IAP Statement on tropical forests and climate change'. URL: www.interacademies.net/Object.File/Master/10/070/Statement_DES1748_IAP%20forests_11.09_P-2-1.pdf

IMS (2009) 'IMS forecasts global pharmaceutical market growth of 4–6% in 2010; predicts 4–7% expansion through 2013' (7 October). URL: www.imshealth.com/portal/site/imshealth/menuitem.a46c6d4df3db4b3d88f611019418c22a?vgnextoid=500e8fabedf24210VgnVCM100000ed152ca2RCRD&cpsextcurrchannel=1

Inauen, C. (2010a) 'Promoting biodiversity through sustainable cocoa sourcing – Experiences from a pilot project in Honduras', Presentation to Helvetas General Assembly, 25 June, Weinfelden. URL: www.biovision.ch/fileadmin/pdf/d/news/ChocolatHalbaNATUR2010.pdf

Inauen, C. (2010b) Personal communication.

Kline & Company (2009) 'Natural personal care products: will the growth continue?' Presentation at the 2009 In-Cosmetics trade fair (April), Munich, Germany.

Kossoy, A. and Ambrosi, P. (2010) *State and Trends of the Carbon Market 2010*, The World Bank, Washington, DC (May).

Laird, S. A. and Wynberg, R. (2005) *The Commercial Use of Biodiversity: An Update on Current Trends in Demand for Access to Genetic Resources and Benefit-Sharing, and Industry Perspectives on ABS Policy and Implementation*. Prepared for the fourth meeting of the Ad Hoc Open-ended Working Group on Access and Benefit-sharing. URL: www.cbd.int/doc/meetings/abs/abswg-04/information/abswg-04-inf-05-en.pdf

Madsen, B., Carroll, N. and Moore Brands, K. (2010) *State of Biodiversity Markets*. URL: www.ecosystemmarketplace.com/documents/acrobat/sbdmr.pdf

Marine Stewardship Council (2009) *Net Benefits: The First Ten Years of MSC Certified Sustainable Fisheries*. URL: www.msc.org/documents/fisheries-factsheets/netbenefits-report/net-benefits-introduction-web.pdf

Mazzanti, R. and Zavettieri, S. (2009) *Summer 2010 Trends: Misty, Mimetic Hues and Weather Proof Materials*, Fiera de Milano Press Office, Milan (16 September). URL: www.mipel.com/files/comstampa96mipel_eng.pdf (last accessed 13 October 2009).

Mulder, I., ten Kate, K. and Scherr, S. (2006) 'Private sector demand in markets for ecosystem services: preliminary findings', Forest Trends. URL: www.fsd.nl/downloadattachment/72341/ 60533/private_sector_demand.pdf

Organic Monitor (2009a) 'Organic Monitor gives 2009 predictions' (30 January). URL: www.organicmonitor.com/r3001.htm

Organic Monitor (2009b) 'Global organic market: Time for organic plus strategies' (29 May). URL: www.organicmonitor.com/r2905.htm

OTA (2009) 'Organic Trade Association releases its 2009 organic industry survey', press release (4 May). URL: www.organicnewsroom.com/2009/05/organic_trade_association_rele_1.html

Plambeck, E.L. and Denend, L. (2008) 'The greening of Wal-Mart', *Stanford Social Innovation Review* (Spring). URL: http://csi.gsb.stanford.edu/greening-wal-mart (last accessed 29 June 2010).

Prescott, J. (2009) 'Buyers pre-empt demand for sustainability', *Ecotextile News* (25 September). URL: www.ecotextile.com/news_details.php?id=10016

PricewaterhouseCoopers and Conservation Finance Alliance (2010) National REDD+ funding frameworks and achieving REDD readiness: Findings from consultation.

State of Victoria Department of Sustainability and Environment (2006) Bushbroker: Native vegetation credit registration and trading.

TEEB in Policy Making (ed.) (2011) *The Economics of Ecosystems and Biodiversity in National and International Policy Making* (ed. P. ten Brink), Earthscan, London.

ten Kate, K. and Laird, S. (1999) *The Commercial Use of Biodiversity: Access to Genetic Resources and Benefit-Sharing*. Commission of the European Communities and Earthscan, London.

Traffic International (2006) *Traffic Bulletin*, vol. 21, no. 1 (July).

TSC (2009) *Partnership for Global Sustainable Tourism Criteria and Sustainable Tourism Stewardship Council Announce Merger to Form Tourism Sustainability Council*. URL: www.sustainabletourismcriteria.org/index.php?option=com_content&task=view&id=266&Itemid=483

UNEP (2008) *Forest Biodiversity*, COP 9 MOP 4, Bonn, Germany.

UNEP and New Energy Finance (2009) *Global Trends in Sustainable Energy Investment 2009*. URL: http://sefi.unep.org/fileadmin/media/sefi/docs/publications/Executive_Summary_2009_EN.pdf

UNEP/CBD/WG-RI/3/INF/13 (2010) *Innovative Financial Mechanisms: Initiating Work on a Green Development Mechanism*. URL: www.cbd.int/wgri3/meeting/Documents.shtml

UNEP/CBD/COP/10/INF/15 (2010) *Innovative Financial Mechanisms: The GDM 2010 Initiative Report*. URL: www.cbd.int/doc/meetings/cop/cop-10/information/cop-10-inf-15-en.pdf

VIETRADE (n.d.) *Vietnamese handicrafts and traditional craft villages 2008–2009*, Hanoi, Vietnam.

VIETRADE (2008) *Trade and Investment of Vietnam: Facts and Figures 2008–2009* (November), Hanoi, Vietnam.

Werf, G.R. van der, Morton, D.C., DeFries, R.S., Olivier, J.G.J., Kasibhatla, P.S., Jackson, R.B., et al. (2009) 'CO_2 emissions from forest loss', *Nature Geoscience*, vol. 2 (November).

Wilhelmsson, D., Malm, T., Thompson, R., Tchou, J., Sarantakos, G., McCormick, N., et al. (eds) (2010) *Greening Blue Energy: Identifying and Managing the Biodiversity Risks and Opportunities of Off-shore Renewable Energy*. URL: http://data.iucn.org/dbtw-wpd/edocs/2010-014.pdf

Willan, B. (2009) 'Twenty trends for sustainability in 2009–10', *Environmental Leader* (24 September). URL: www.environmentalleader.com/2009/09/24/twenty-trends-for-sustainabilityin-2009-10/

Wille, C. (2009) Rainforest Alliance – Sustainable Agriculture Network Presentation at the Sustainability Conference, Nuremberg (February).

World Health Organization (2003) *WHO Guidelines on Good Agricultural and Collection Practices (GACP) for Medicinal Plants*, WHO, Geneva.

Websites

www.bamboosushipdx.com (last accessed 9 July 2010).
www.eartheasy.com (last accessed 9 July 2010).
www.ecoclub.com (last accessed 9 July 2010).
www.ecoenterprisesfund.com (last accessed 9 July 2010).
www.ecotourism.org (last accessed 9 July 2010).
www.fairwild.org (last accessed 9 July 2010).
www.gdm.earthmind.net (last accessed 9 July 2010).
www.lohas.com (last accessed 9 July 2010).
www.msc.org (last accessed 9 July 2010).
www.missionmarkets.com (last accessed 9 July 2010).

Chapter 6

Business, Biodiversity and Development

Editor
Linda Hwang (BSR)

Contributing authors
Suhel al-Janabi (GIZ), Andreas Drews (GIZ), Joshua Bishop (IUCN), Matthew Lynch (WBCSD)

Contents

Key messages

Economic and social development generally involves more consumption and open markets, both highly correlated with business development but also often associated with biodiversity loss and ecosystem decline. The challenge is to reinforce economic development strategies that are ecologically sustainable, socially equitable and also good for business.

There is a strong business case for companies to engage in social development. Contributing to the development of regional and national economies enhances a company's operational capability, opens new markets and reduces potential negative impacts to reputation. Companies that take a strategic approach to social investment can align both corporate and community needs.

There are potential synergies between business, biodiversity conservation and poverty reduction, but these are not realized automatically. Biodiversity and ecosystem services are not routinely considered in decisions related to corporate social investment programmes. Many companies have programmes that support biodiversity conservation and separate programmes that support local economic development. In some cases these programmes are in conflict, although a few companies have found ways to combine biodiversity and ecosystems with their social programmes.

There are many opportunities for corporate action, once the linkages between poverty and biodiversity and ecosystem services are understood. There are increasing examples of business aligning their contributions to local economic and social development with conservation or enhancement of biodiversity and ecosystem services (BES) in ways that generate both business and societal value. The most significant and scalable actions are those associated with core business activities (e.g. product and services, procurement), but there are also important opportunities through strategic social investment, institutional capacity building, collective action and multi-stakeholder partnerships.

Good business for the company must also be good business for the local community. Gaining the 'social licence to operate' as well as government goodwill has become a standard requirement in the business world. This includes making a visible contribution to local environmental quality. Business can help meet both development and environmental objectives, for example by engaging with other sectors operating in the region and/or country to enhance development initiatives and scale up pilot initiatives, as well as by undertaking direct conservation actions.

6.1 Introduction

Sustainable development, according to one of the most widely accepted definitions, is development that meets the needs of the present without compromising the ability of future generations to meet their own needs (Brundtland 1987). Poverty eradication, the reduction of unsustainable patterns of production and consumption, and the protection and management of natural resources are frequently cited as both overarching objectives and essential requirements for sustainable development.

Reducing poverty and achieving sustainable development is made more difficult as biodiversity and ecosystem services are degraded. As noted in Chapter 2 and more fully documented in a range of recent reports, people have made unprecedented changes to ecosystems in recent decades to meet growing demand for food, fresh water, fibre and energy. The quality of life of billions of people has improved, although extreme poverty remains a major challenge in many regions of the world – particularly in sub-Saharan Africa and south Asia – with 1.4 billion people globally still living on less than US$1.25 per day (United Nations 2010).

At the same time, the environmental changes associated with economic growth have often weakened the ability of nature to deliver essential services, such as air and water purification, protection from natural disasters and climate stability. The Millennium Ecosystem Assessment (MA), among many other analyses, has documented how the loss of ecosystem services can undermine efforts to reduce poverty, hunger and disease (MA 2005). Biodiversity and ecosystem services are of particular importance for poverty reduction because so many of the world's poorest people rely directly on these assets and services for their well-being.

This chapter explores the relationship between business, biodiversity and social development, highlighting the efforts made by some companies to align their contributions to local economic and social development with the conservation or enhancement of BES, in ways that generate both business and societal value. The private sector is increasingly recognized as an essential actor in development and poverty reduction, and companies themselves are increasingly seeing the business opportunities associated with engaging in development challenges. However, the complex linkages between poverty and BES require companies to adopt a more integrated perspective if they are to make effective and efficient contributions to sustainable development. This chapter presents a range of opportunities for business action, while also highlighting some of the challenges that business face when attempting to integrate BES with core business and social investment activities.

6.2 The business contribution to development

The role of the private sector in development and in poverty reduction is well established. Encouraging greater private sector activity as a primary mechanism for economic and social development is a fundamental concern of developing country governments and many governmental and multilateral development agencies. As summarized in the World Bank's 2005 *World Development Report*:

> Private firms are at the heart of the development process. Driven by the quest for profits, firms of all types ... invest in new ideas and new facilities that strengthen the foundation of economic growth and prosperity. They provide more than 90 percent of jobs – creating

> opportunities for people to apply their talents and improve their situations. They provide goods and services needed to sustain life and improve living standards. They are also the main source of tax revenues. ... Firms are thus central actors in the quest for growth and poverty reduction. (World Bank 2005a: 1)

While the contribution of business to development is not in doubt, there remains a question about how much business can and should be expected to deliver.

6.2.1 The scope and limits of business responsibility

Development assistance strategies have long relied on the private sector as an engine of economic growth. Business contributes to poverty reduction in many ways, through job and wealth creation, transfer of technology and skills, and the efficient supply of affordable goods and services to a range of final consumers and other businesses, including some of the poorest people at the 'base of the pyramid'. (UNDP 2004).[1] The challenge for companies is to determine what role they can play in creating new economic opportunities, so that individuals can develop their own solutions to generate more income and realize other opportunities that improve their (and others') quality of life.

While the private sector is a key engine of economic growth, there are limits to what business can do. In particular, business cannot replace governments in certain key aspects of the development process. Indeed, the quality and effectiveness of governments (i.e. whether they are capable of helping business grow, delivering services to their citizens, and are accountable and responsive to them) 'is the single most important factor that determines whether or not successful development takes place' (DFID 2006: ix).

Furthermore, the relationship between private sector activity, development and poverty reduction is not straightforward. To begin with, the poor may not share equally in market-led growth. In some situations, they may end up worse off either absolutely or relatively, due to changes in the distribution of income and wealth that result from market-led growth or globalization. To what extent such adverse social outcomes are an inevitable result of market-led growth or the result of imperfect competition and the persistence of market barriers is a matter of continuing debate (Rajan and Zingales 2006; Milanovic 2006).

6.2.2 The business case for development

Despite these limitations and uncertainties, both local and international firms with operations in low-income or 'emerging' economies are increasingly engaging actively with the development challenge. At a basic level, there is growing recognition that responsible business practices and standards are necessary 'to manage risks, minimize negative externalities and impacts, and protect existing market and social value' (Nelson 2006: 2). This 'responsible business' approach is embodied in the United Nations Global Compact, for example, in which some 7,700 companies and other stakeholders have committed to align their strategies and operations with universally accepted principles in the areas of human rights, labour, environment and anti-corruption (UN Global Compact 2010).

At the same time, some firms are adopting a more strategic approach to identify 'win–win' opportunities that can make significant contributions to development while also generating commercial returns or other benefits for the firm. The business case for these kinds of strategic actions varies with the company, sector and local context; however, commonly noted drivers include:

- *New markets* – there is growing recognition that the world's 4 billion poorest consumers represent a significant under-served market for business (Prahalad 2006). Firms that can reach these 'base of the pyramid' consumers by offering appropriate and affordable products and services can reap significant commercial benefits, as demonstrated by the examples in Box 6.1. In addition, the emerging middle class in India, China, Brazil and other developing economies is increasingly attracting companies in consumer goods and manufacturing industries.
- *Social licence to operate* – establishing strong economic and social linkages with surrounding communities can be a key strategy for building positive relationships and mitigating the risks of stakeholder dissatisfaction, especially for firms whose operations affect the physical environment or impinge on social and cultural values.
- *Brand and reputation* – firms can gain reputational benefits by making positive contributions to development, including consumer goods companies as well as extractive industries seeking to secure access to resources. Business action on development can also contribute to staff morale, recruitment and retention.
- *Supply chain management* – many companies rely on products or services produced by relatively poor producers, especially in the agriculture sector. Investing in the capabilities of these producers can reduce costs and improve the quality and reliability of supply chains, with the additional benefit of increasing producer incomes and reducing their vulnerability.

These and other drivers of business action on development are summarized in the concept of 'inclusive business models' (WBCSD 2005), which 'include the poor on the demand side as clients and customers, and on the supply side as employees, producers and business owners at various points in the value chain. They build bridges between business and the poor for mutual benefit' (UNDP 2008).

Box 6.1 Examples of inclusive business models.

Inclusive business models have been developed in a range of sectors and locations. Prominent examples from four different sectors are presented below.

- *Telecommunications/financial services* – Vodafone M-PESA is a mobile-phone-based service developed originally in Kenya by Safaricom (Vodafone), which allows customers without access to bank accounts to perform basic banking transactions. M-PESA currently has more than 9 million customers and has been extended to several other countries (Vodafone 2010; WBCSD 2008).
- *Microinsurance* – Allianz, in collaboration with partners including UNDP, GIZ, CARE International, and numerous microfinance initiatives and cooperatives, is providing

life, property and health insurance services to 3.8 million poor customers in India, Indonesia, Africa and Latin America (Emergia Institute and Allianz Group 2010).

- *Water supply* – Manila Water has an innovative programme called Tubig Para Sa Barangay (TPSB, or 'Water for Poor Communities'), designed to reach low-income communities and 'based on a clear business case: underserved, low-income households demonstrate a willingness to pay for safe, reliable water and connecting them means reaching new markets while reducing costs from inefficiencies and illegal connections' (Jenkins and Ishikawa 2010: 36). A range of models are used for service provision, but most are based on metered, small-group connections. Since the programme's start, TPSB has expanded reliable and safe drinking water services to 1.6 million people, significantly reducing the costs paid by poor consumers for water and greatly increasing the profitability of the concession (Jenkins and Ishikawa 2010).
- *Mining/access to finance* – since 1989, the mining group Anglo American has operated an enterprise investment and development fund in South Africa called Anglo Zimele, which empowers black entrepreneurs through the creation and transformation of small and medium enterprises (Anglo American and IFC 2008). The fund has provided financing to over 150 local ventures through a combination of small equity stakes and loans, as well as technical assistance and easier access to procurement. Since 1997, the survival rate of firms supported by Anglo Zimele has been triple the national average (Anglo American 2005).

6.3 Biodiversity and ecosystem services and the business role in development

Business contributions to development and poverty reduction usually focus on promoting local economic and social development. However, it is essential that the linkages between poverty and BES are considered when designing such initiatives, if they are to make genuinely sustainable contributions to human well-being. Increased economic activity does not automatically result in less poverty, a better environment, greater social equity or a higher quality of life for local populations. Indeed, where economic activity results in adverse impacts on BES, it is often the poorest and most vulnerable groups that suffer most. Conversely, more holistic consideration of the linkages between poverty and BES can enhance the opportunities for businesses to generate both commercial and societal value. These linkages and opportunities are explored in more detail in the following pages.

6.3.1 The links between biodiversity, ecosystem services and poverty

Ecosystem services are fundamentally important for human well-being and development (MA 2005). Billions of people directly depend on the health of natural systems for their livelihoods. The MA also shows how addressing the development needs of the poor can reduce some of the pressures on ecosystems.

The complex linkages between poverty, human population change and BES are sometimes referred to as the 'poverty–environment nexus' (Pearce 2005). A better understanding of these linkages can help firms identify opportunities and risks, and integrate their actions to address poverty and/or the decline in BES. Key features of the poverty–environment nexus are outlined below.

Many poor people depend directly and disproportionately on biodiversity and/or ecosystem services for their livelihoods. Three-quarters of the world's poorest people – about 1.2 billion individuals who live on less than one US dollar a day – live in rural areas, and their survival depends in large part on primary production (Cervantes-Godoy and Dewbre 2010). Poor rural households in particular rely on both cash income and direct subsistence from the use of natural resources, such as small-scale farming, fishing, hunting and the collection of firewood or other natural products (WRI 2005). A large proportion of the rural poor rely heavily on forests for their livelihoods and forest-related income provides a significant share (around 20 per cent on average) of total household income in poorer regions (Pearce 2005).

The modest purchasing power of poor households leaves them less able to obtain substitutes for local natural resources, if these become scarce or off limits. As a result, they remain highly dependent on the integrity of the local environment for the supply of food, water, energy, shelter and other basic needs. From the perspective of rural communities, it is the local values of biodiversity that are most important: the distribution and abundance of wild species, the range of crop plants and livestock available, and the diversity of ecosystems that are easily accessible (Ash and Jenkins 2007).

Environmental assets make up a significant share of the wealth of the poor but are prone to rapid depreciation, unless they are managed and regenerated. Natural-resource-dependent communities are particularly vulnerable in regions experiencing rapid population growth. With 'few assets, low-quality assets and lack of access to technology to make their assets more productive, poor households and communities may have incomes that are too low to generate re-investable surpluses for maintaining, much less expanding, their [natural] asset base' (Pearce 2005: 10). The result can be a vicious cycle of environmental degradation, increasing vulnerability and poverty.

As noted above, the quality and effectiveness of governance is a critical determinant of development outcomes. Unfortunately, in many countries, weak governance is compounded by inadequate or inappropriate reflection of the linkages between BES and poverty in policy and decision making (see Box 6.2). Insecure resource rights and other disincentives to wise management and use of natural resources often contribute significantly to their degradation (Pearce 2005).

Box 6.2 Poverty and ecosystem degradation

Despite the seemingly obvious links between ecosystem services and poverty, the reliance of the poor on ecosystem services is rarely measured and is thus typically overlooked in national statistics, poverty assessments and land-use and natural resource management decisions. In particular, the patterns of winners and losers associated with ecosystem change, and their impact on the chronically poor and women in particular, has been given little consideration. Such inattention can lead to inappropriate strategies that ignore the role of environment in poverty reduction, possibly leading to further marginalisation of the poorest sectors of society and increased pressure on ecosystems.

Source: Shackleton et al. (2008)

6.3.2 Opportunities and challenges in the poverty–environment nexus

Experience shows that there are potential synergies between business, conservation and poverty reduction. Examples include private sector participation in water supply and sanitation (Johnstone and Wood 2001), agribusiness development (McNeely and Scherr 2002), forest product markets (Macqueen 2008) and payments for ecosystem services (Pagiola et al. 2005). However, these synergies are not realized automatically.

Business has a major role to play in development and poverty reduction, working alongside governments and NGOs. Business can also make significant contributions to conserving biodiversity and supplying ecosystem services, as described in other chapters in this book. The argument here is that the close linkages between poverty and BES demand more integrated consideration of these actions. For example, while business actions that lead directly or indirectly to adverse impacts on BES can have significant negative repercussions for poor communities, business efforts to support improved management and governance of BES, or to increase the productivity and sustainability of natural resource use by poor communities, can make a significant contribution to poverty reduction.

The question for business is how to align business success with poverty reduction, business value with BES, and BES with human development. The following sections examine the opportunities (Section 6.4) and the challenges (Section 6.5) for business in pursuing such an agenda.

6.4 Linking biodiversity, ecosystem services and development

This section outlines opportunities for business to align their contributions to local economic and social development with the conservation or enhancement of BES in ways that generate both business and societal value. The focus is on core business activities, but we also acknowledge the opportunities associated with other corporate actions – including strategic social investment and institutional engagement, as well as collective action and multi-stakeholder partnerships. This section builds on the analysis of BES-related opportunities for business outlined in Chapter 5, many of which are highly relevant in a development context.

An emerging consensus suggests several ways that business can benefit from engaging in development (adapted from Nelson 2006):

- core business operations;
- strategic social investment;
- institutional capacity building;
- collective action and multi-stakeholder partnerships; and
- benefit sharing.

Each of these is discussed in turn below.

6.4.1 Core business operations

While firms historically engaged with social and community issues on a philanthropic basis, there is now widespread acknowledgement that contributions through core business systems and processes are potentially much larger, as well as being more sustainable and scalable (WBCSD 2005).

New products and services

A key opportunity is the development of new products or services (possibly using inclusive business models) that contribute to both social development and the conservation or enhancement of BES. The development of new business models may involve demand-side interventions, such as educating consumers, as well as supply-side actions or investments in production. Box 6.3, for example, describes the development and commercialization of a new edible oil by Unilever, working in cooperation with several partners in Africa. This approach is further demonstrated in many of the BES-related products and services described in Chapter 5, in sectors such as tourism, cosmetics, garments and finance.

Box 6.3 Unilever: Development of *Allanblackia* oil

Unilever is one of the world's largest suppliers of fast-moving (as opposed to durable) consumer goods, with products on sale in more than 170 countries. Unilever has made sustainability a key part of its branding strategy and has committed to explore opportunities to increase sourcing from smallholder farmers while also ensuring security of supply. This is in line with a wider strategy of 'shared value creation' with the communities in which the company operates.

As part of a commitment to working with smallholder farmers, Unilever is developing new commercial uses of *Allanblackia* trees, which occur naturally in the wet tropical forests of Africa. This endemic tree species produces a large fruit pod, containing seeds that are rich in edible oil with many useful properties. In particular, spreads containing *Allanblackia* oil, such as the well-known brands Flora and Becel, remain stable at room temperature but melt quickly upon eating. In 2008, the European Commission approved *Allanblackia* oil for use in spreads and it is now an ingredient in several Unilever products on sale in Europe.

The development of *Allanblackia* on a commercial basis has provided an opportunity for the company to support more sustainable agricultural practices with local farmers, while also ensuring the conservation of threatened ecosystems in growing regions. Unilever are currently working with over ten thousand smallholder farmers in several countries to expand *Allanblackia* production.

Sources: Unilever (2010a, b)

Procurement standards

Rigorously applied supply chain standards can support the integration of BES and development through a company's procurement of goods and services from developing regions. Many of the prominent certification schemes for consumer products presented in Chapter 5 (e.g. Rainforest Alliance, Fairtrade) include requirements related to both BES and community impacts. Such approaches can also be developed and implemented at an individual company level, as is illustrated by the Starbucks' CAFE Practices (Box 6.4).

Box 6.4 Sustainable sourcing at Starbucks Coffee

Starbucks has established buying guidelines for ethical sourcing in partnership with Conservation International (CI). Called the Coffee and Farmer Equity (CAFE) Practices, the guidelines set measurable standards targeted to the adoption of socio-economic and environmental best practices within Starbucks' supply chain. Of the 367 million pounds of coffee bought by Starbucks in fiscal year 2009, 299 million pounds (81 per cent) were sourced from CAFE approved suppliers. In addition, Starbucks is the largest purchaser of Fair Trade-certified coffee (40 million pounds per year), as well as a major buyer of organic coffee (14 million pounds per year).

The CAFE Practices focus on four issues:

- Product quality.
- Economic accountability – evidence of payments from suppliers demonstrates how much revenue goes to farmers.
- Social responsibility – measures have been put in place to ensure safe, fair and human working conditions.
- Environmental leadership – agricultural practices are promoted that conserve biodiversity, reduce the use of agrochemicals, improve waste management, protect water quality, and conserve water and energy.

Implementation of the CAFE Practices focused initially on suppliers from 19 countries on four continents, overlapping 18 of the world's most biologically rich regions. In addition to development benefits for coffee-farming communities, a 2007 survey found that 99 per cent of farmers participating in the CAFE programme had managed to avoid converting natural forest to coffee production over the preceding three years.

Sources: Starbucks (2010); CI (n.d., 2008)

Supply chain development

Many companies purchase raw materials or semi-finished products from relatively poor producers in developing countries. There are often strong commercial reasons to improve the quality and reliability of these inputs, which may be achieved by helping farmers increase productivity (e.g. by improving access to knowledge, skills, technology and finance). Productivity improvements can enhance farmer income while also supporting the adoption of more sustainable farming practices and use of natural resources. Box 6.5 provides an example of the synergies between business, environment and development that can be achieved by investing in farmer productivity.

Box 6.5 Jain Irrigation Systems, water management and farmer incomes in India

The management of water resources is a major environmental and development challenge in India, as in many other countries. Current levels of surface and groundwater abstraction in many parts of the country are considered unsustainable (World Bank 2005b).

This has significant negative impacts on a range of aquatic and terrestrial ecosystems, and puts at risk the well-being of millions of poor farmers, who rely heavily on freshwater resources for their livelihoods.

Jain Irrigation Systems Ltd (JISL), based in India, has developed a business model that responds to these challenges. JISL is the largest manufacturer of efficient irrigation systems worldwide and a major processor of fruits and vegetables. The company provides farmers with micro-irrigation systems (MIS), seeds and other inputs to produce more and better crops. The company also purchases fruits and vegetables through its food division, which processes them and sells them to both export and domestic markets.

The use of JISL drip and sprinkler irrigation, as opposed to traditional flood irrigation techniques, has reduced farmers' water use by an estimated 500 million cubic metres per year. As well as MIS, farmers under contract with JISL benefit from high-quality seeds, financing for inputs, agronomic support and an assured market for their crops, generating an additional US$300–400 per acre in revenue, compared with conventional growing practices. Independent farmers who use only the JISL micro-irrigation system have also increased efficiency and reduced their vulnerability to unreliable or over-exploited water resources. Net incomes have increased by US$100 to US$1,000 per acre, depending upon the crop, with the investment cost of MIS typically recouped in less than one year.

Source: Jenkins and Ishikawa (2010)

Diversification of local economic opportunity

Companies in many sectors have found value by promoting local economic participation in their activities, including direct employment as well as local suppliers and contractors. In poor rural areas this often involves significant efforts from the company (e.g. provision of training and financing) to strengthen local capabilities. In addition to the direct development benefits of such relationships, they may help to broaden the range of livelihood options available to local people and thus reduce the need to exploit increasingly scarce natural resources.

6.4.2 Strategic social investment

Many companies are integrating community development and BES objectives in their social investment programmes, moving away from pure philanthropy to a more strategic approach to social investment. This often involves efforts to maximize the value of corporate contributions (both financial and non-financial) to local development, while also aligning these investments with the company's core competencies and its wider sustainability and commercial objectives.

Some companies in the extractive sectors, for example, have started to think of their community or social investments not only as a way of 'doing good' but also as a means to create an operating environment that is conducive to their business. By investing strategically in local communities, companies can not only help to develop a skilled workforce and more competitive local suppliers, but can also support the expansion of economic opportunities and secure environmental assets on which local communities depend.

A strategic approach to social investment is often more difficult to implement than conventional corporate philanthropy but, if done well, its effects can be longer lasting and more in line with both company and community needs (IFC and BSR 2008). Social investment programmes typically involve working in partnership with community-based organizations and NGOs. Key elements of these programmes may include the following:

Local capacity building

This supports the development of the local community's knowledge, skills and institutional capacity to manage natural resources as well as other challenges. Supporting communities to build and maintain effective local institutions can improve their ability to seize new opportunities while also reducing vulnerability to external shocks. In many cases, companies may find it most effective to build upon existing formal or informal institutional structures in order to enable local communities to improve the management of natural resources.

Direct investment in local environmental assets

Companies can assist poor communities by investing directly in the conservation or rehabilitation of important natural resources (see Box 6.6).

Box 6.6 Linking biodiversity with community social drivers

In 2004, Comsur operated the Don Mario gold mine in Bolivia, located in a rare dry forest ecosystem of global significance that is home to many endemic species and nearly 100 threatened or endangered species. Comsur's biodiversity programme included fencing off its operational areas, to prevent access by animals, closing roads to prevent illegal logging and a 12,000-tree reforestation programme. At the same time, Comsur provided employment opportunities to surrounding communities as well as financial assistance for educational, sanitation and infrastructure projects. The mine also purchased a range of agricultural products from local communities. Through surveys and public meetings in the different communities and in partnership with government authorities in San Juan, Comsur analysed the needs of each community and developed programmes to respond to these.

In another example, Peru LNG worked closely with the National Council for the Environment to build partnerships with local NGOs and other initiatives that aligned with the company's goals for biodiversity conservation (IPIECA 2006). By engaging local stakeholders with both biodiversity conservation expertise and long-standing relationships with communities, Peru LNG was able to build trust and establish goodwill with local communities.

Sources: IFC (2004); IPIECA (2006)

Strengthening sustainable livelihoods

Wherever there is primary production and associated goods and services, companies have an opportunity to strengthen the ability of people to participate in economic activity, to enhance the productivity of those activities or to improve the management of the ecological assets that underpin those activities. All this can support the development of livelihoods based on the sustainable use of local natural resources, which may offer more opportunities than direct participation in the company's operations. For example, the Alcoa Juruti mine in the Amazon region of Brazil has invested in local economic diversification and the development of alternative livelihood opportunities for communities affected by the project. Many of these projects are focused on sustainable use of non-timber forest resources. Such investments can help broaden the sustainable development outcomes associated with a company's activities, enhancing its social licence to operate and reducing the risks of a 'boom–bust' cycle, which all too often results from extractive industry developments in remote regions (Alcoa 2010).

6.4.3 Institutional capacity building

Good governance is fundamental to development. Private sector engagement in this area can be complex but there is an emerging consensus that companies need to advocate and reinforce institutional strengthening and improved governance, in order to support local development and improved environmental management.

In many developing countries, the capacity of local authorities and government agencies to manage natural resources is limited. Improving local capacity is often in the interests of companies operating in such contexts, as it reduces pressures on companies to 'substitute' for missing or weak government services, while ensuring adequate management and services in areas around a company's operations but outside of its direct control. A growing number of companies are combining advocacy for improved natural resource management by government agencies with support for government capacity building, typically in partnership with other stakeholders. For example, BP Indonesia has worked with UNDP, USAID and the UK Department for International Development to support the development of government environmental management capacity, as part of the Tangguh LNG project in West Papua, Indonesia (BP 2010).

Where a company has significant expertise based in a particular area, there may be opportunities to link with and support the development of local government capacity (e.g. through formal and/or on-the-job training). The resulting improvements in government capacity can help protect the natural resources on which both the business and poor people in the area depend.

Similarly, it is often in companies' interest to support the development of effective local systems for defining and enforcing access and property rights over natural resources. If managed appropriately, these systems can have significant positive spill-over effects for poor resource users and help improve the management of natural resources.

6.4.4 Collective action and multi-stakeholder partnerships

There are many development challenges and opportunities that 'directly affect a company's profitability or operating environment, [but] which it is unable to address either effectively, legitimately, or to a sufficient scale on its own' (Nelson 2006: 11). In such circumstances there may be grounds for collective corporate action, cross-sector engagement and/or multi-stakeholder partnerships.

Given the complexity, scale and many linkages between development and BES challenges, this is becoming an important area of business action. One example is the development of industry-level product or certification standards, which are typically developed in partnership with a range of stakeholders, such as the Roundtable on Sustainable Biofuels (RSB) and Roundtable on Sustainable Palm Oil (RSPO). The RSPO definition of sustainability includes biodiversity conservation, reflecting the fact that most of the world's palm oil is produced in areas rich in biodiversity. In addition to environmental concerns, the RSB 'Principles and Criteria for Sustainable Biofuel Production' also define performance-related principles relating to rural and social development, human and labour rights, and land rights (RSB 2009).

6.4.5 Benefit sharing

Many industries rely on genetic material or information derived from biodiversity-rich developing countries. Moreover, the industrial potential of particular species is often based on the traditional knowledge of communities that have used these resources for centuries. However, while the profits from exploiting wild genetic resources can be significant, the benefits arising from such uses have historically not been shared equitably with the countries and communities of origin. Ensuring that appropriate benefit-sharing regimes are in place can help increase the value that accrues to local communities while also building the social licence of business to access other resources in the future.

The value of wild genetic resources as inputs to the pharmaceutical industry has been noted in preceding chapters (and in several other TEEB publications, notably TEEB in Policy Making 2011). Other sectors that rely on genetic resources and/or naturally occurring biochemical compounds include biotechnology, seed production, animal breeding, crop protection, horticulture, cosmetics, fragrance, botanicals, and the food and beverage industries.

Until recently, developing countries had no means to ensure that the utilization of their resources, traditional knowledge and cultural practices would be recompensed. Hence one of the three objectives of the Convention on Biological Diversity (CBD) is to regulate 'access to genetic resources and ensure the fair and equitable sharing of benefits arising out of their utilization' (www.cbd.int/abs/regime.shtml). The principles laid down in Article 15 of the CBD on access and benefit sharing (ABS) provide significant opportunities for poverty alleviation and sustainable development.

Recently, considerable effort has been devoted to negotiating and establishing ABS in international and national law and in voluntary guidelines. An important milestone was the development of the voluntary *Bonn Guidelines* on access and benefit sharing, which addressed issues such as 'mutually agreed terms' and 'prior informed consent' for the use of genetic resources (SCBD 2002). Despite this agreement, at the World Summit on Sustainable Development (WSSD) in 2002, the world's governments concluded that voluntary guidelines were not able to address ongoing problems of 'mis-appropriation' of genetic resources. The WSSD called on the CBD to develop 'an international regime to promote and safeguard the fair and equitable sharing of benefits arising out of the utilization of genetic resources' (see Chapter IV, subsection (o) of the WSSD Plan of Implementation). Finally, at the tenth Conference of the Parties to the CBD in late 2010, these efforts culminated in the agreement of the Nagoya Protocol on Access to Genetic Resources and the Fair and Equitable Sharing of Benefits Arising from their Utilization.

Reaching an international agreement on ABS was not easy, in part because of the many issues that need to be addressed. In general, all of the ABS frameworks and initiatives listed above aim to:

- support poverty alleviation and nature conservation;
- support capacity development by transferring technologies, knowledge and skills;
- enhance social development especially in countries of origin; and
- ensure accountability and good governance at all levels.

While the ultimate impacts of the Nagoya Protocol remain to be seen, examples of how it might work include some recent voluntary agreements that combine access to genetic resource with biodiversity conservation and benefit sharing (Box 6.7).

Box 6.7 Novartis Coartem anti-malaria treatment: A win–win–win for biodiversity, health and poverty reduction?

In 1994 the Swiss-based Novartis healthcare multinational, together with Chinese partners, started to develop the combination drug Coartem, which is currently the world's leading Artemisinin-based combination therapy for the treatment of malaria. The product is based on extracts of the *Artemisia annua* plant, which originates in China and other southeast Asian countries. Access to the resource was granted by the Chinese government and is now secured by several suppliers, mainly in China. *Artemisia* is currently mainly cultivated, using improved seed material, rather than collected from the wild, in order to meet high demand and reduce the pressure on the species' natural habitat.

The active components Artemether and Lumefantrine were known to Chinese traditional medicine and further developed and refined by Novartis research laboratories.[2] Coartem patents have been granted in roughly 50 countries (international application number PCT/EP1999/004355) and are co-owned by Novartis and the Institute of Microbiology and Epidemiology and the Academy of Military Medical Sciences of the Government of China.

Benefit sharing by Novartis includes scientific training, technology transfer (e.g. laboratory equipment), about US$150–160 million for the supply of raw material from Artemisia producers, plus additional royalties and other payments to the Chinese scientific partners.

In 2001 Novartis entered into a Memorandum of Understanding with the World Health Organization to provide Coartem to public agencies in malaria-endemic developing countries at not-for-profit prices. On the basis of this agreement, the number of treatments provided increased from 4 million in 12 countries (in 2004) to 84 million in over 60 countries (in 2009). It is estimated that about 800,000 lives were saved as a result of the delivery of 320 million Coartem treatments. The average sale price of Coartem to authorized public buyers in 2010 is less than one US dollar (i.e. 76 US cents for one adult treatment or 36 cents for a child). Novartis estimates the full economic value of these treatments at about US$1 billion, if Coartem had been supplied on normal 'for-profit' basis.[3]

Source: al-Janabi and Drews (2010)

6.5 Risks and challenges

While there are many opportunities to combine business success with social development and biodiversity conservation, there are also risks and challenges that must be acknowledged. This section highlights some of the issues, shows how even well-intentioned business action can have adverse unintended consequences and suggests some precautions that companies can take.

6.5.1 System understanding and measurement issues

Decisions about natural resource conservation and restoration require a good understanding of the services provided by ecosystems and of how these systems respond to human interventions or to natural phenomena, supported by reliable metrics. Similarly, decisions about business expansion and economic development require a good understanding of how human communities rely on biodiversity and ecosystems services. This understanding, which is lacking on many levels in most firms includes the following aspects:

- *The ecological processes that produce ecosystem services* – Understanding biophysical processes can help predict the effects of human actions and other natural phenomena, both beneficial and detrimental, on the quantity and quality of ecosystem services and is essential to assess management alternatives.
- *Spatial and temporal dimensions of ecosystem services* – Decisions made today or locally have the potential to affect the availability and quality of natural resources and ecosystem services elsewhere or at some point in the future. For example, Eucalyptus plantations in southern India provide wood pulp and tannin, but affect downstream hydropower projects due to reduced water yield from forested catchments (Rodríguez et al. 2006). All too often, business managers (like other decision makers) are not aware of the wider or long-term ramifications of their decisions.
- *The complex linkages between poverty and the environment* – Many firms find it challenging to predict or monitor the impacts of their actions on local communities. While there are often clear business benefits from supporting local economic and social development (ODI and EAP 2007; IFC and BSR 2008), including reduced risk and increased social licence to operate, the real effectiveness of such support can be difficult to determine due to the influence of other confounding factors.

Organizations such as the International Petroleum Industry Environment Conservation Association (IPIECA), the International Finance Corporation (IFC) and the Centre for Good Governance provide guidance and tools on how to conduct social impact assessments, which can help firms identify the positive and negative impacts of their activities. However, these organizations also acknowledge the complexity of assessing the social and environmental dimensions of business activities (IPIECA and OGP 2002). Corporate social development programmes may include support for alternative livelihood strategies as a means to reduce pressure on natural resources, for example, but evidence suggests that this does not always occur even if successful alternatives are provided (Sievanen et al. 2005). In addition, the difficulties of implementation may undermine the effectiveness of such initiatives (Wells et al. 2007).

6.5.2 Conflicts between development and biodiversity conservation

In some cases, there may be conflicts between corporate programmes focused on human development and those aimed at environmental sustainability. This is one of the more challenging aspects of integrating BES in business.

Certain strategies for reducing poverty can increase pressures on biodiversity and compromise the capacity of ecosystems to produce and maintain services over the long run (Box 6.8). Business investments intended to expand the supply of tradable commodities (e.g. food production) can degrade biodiversity, reduce supplies of other resources (e.g. fresh water, fish, wildlife) and may jeopardize important sources of income to certain segments of the local population. For example, logging natural forests for their timber or clearing them for mining will tend to decrease carbon sequestration, flood control and biodiversity values. This can undermine future provisioning services and increase the vulnerability of local populations to environmental change.

Box 6.8 Poverty reduction and biodiversity conservation: A trade-off?

The periphery of the Volcano National Park in Rwanda is famous for its mountain gorilla population, which supports a lucrative tourist trade. However, the area also has excellent soils, which are likely to increase in value as the use of fertilizer and improved potato varieties double and even triple potential crop yields. Concurrently, export markets are becoming more accessible with the development of better roads and communications networks. There is a strong chance that economic development and poverty-reduction efforts will increase the risk of agricultural encroachment on protected areas around the volcano, due to rising land values and potential profits from farming.

Source: Adapted from Mellor (2002)

At the same time, corporate support for biodiversity conservation and rehabilitation can have unexpected adverse impacts on both local communities and ecosystems that must be taken into account (Box 6.9). In short, companies must take care to ensure that interventions designed to address biodiversity conservation and social development are coherent or at least well coordinated, while recognizing that it can be challenging to achieve both social and environmental objectives simultaneously.

Box 6.9 Aligning social and environmental impacts: Rio Tinto in Madagascar

Rio Tinto has a policy goal of net positive impact (NPI) on biodiversity in its operations. The company aims to achieve NPI by combining state-of-the-art avoidance, mitigation and ecosystem restoration with biodiversity offsets and other conservation actions. In Madagascar, as part of its offset strategy, the company is contemplating potential support for the conservation of some 60,000 hectares of lowland rainforest, as a form of partial compensation for the unavoidable residual impacts of its mining operations in the region.

In this case, the area to be conserved and the resulting biodiversity benefits are thought to meet and possibly exceed the conservation gains required to compensate for the residual biodiversity impacts of the company's mining operation. A study was commissioned to estimate the potential monetary value of these biodiversity benefits. The study examined the costs of conservation, including both up-front investment as well as the management costs of protected areas, together with the opportunity costs that local people bear when they lose access to land that had historically provided food and cash income in lean periods as well as a resource or reserve for future agricultural expansion.

The ecosystem benefits considered included wildlife habitat (valued at US$2.9 million in net present value terms over 30 years, using a 5 per cent discount rate), hydrological regulation (US$470,000) and carbon storage (US$26.8 million). The study concluded that there are significant net economic benefits associated with conservation (about US$17.3 million net of all costs). However, while many of these benefits accrue globally (e.g. wildlife conservation, carbon storage), the costs of conservation are mainly borne by local communities, whose access to forest resources would be restricted to some extent under a conservation scenario.

The study underscored the need for, and the potential scale of, compensation for the opportunity costs incurred by local populations, for example through payments for ecosystem services (PES). The ecosystem values that are easiest to capture are generally those associated with carbon. While the value of carbon storage is significant in this case, local communities would need to receive at least one-quarter of the potential revenues from Reducing Emissions from Deforestation and Forest Degradation (REDD) in order not to be disadvantaged by conservation.

Source: Olsen and Anstee (2011)

6.5.3 Project-induced in-migration

Large-scale land-use projects can dramatically change social conditions around the project site. Early project documents typically outline the potential benefits that local people in the project area might expect. These benefits generally include measures to mitigate adverse impacts, compensation for loss of resources or environmental damage, promises of participation in the project through employment or purchase of goods and services, and project-related community development programmes. Such promises often raise local expectations of a project's potential to improve people's lives, particularly in remote regions that have not previously benefited from national development programmes (USAID 2010; Banks 2009).

In many cases, the economic opportunities associated with large-scale projects can generate significant in-migration of people from elsewhere, with dramatic impacts on local social structures and relationships (Banks 2009; IFC 2009). An example is provided in Box 6.10.

Box 6.10 Freeport Indonesia and the impacts of in-migration

Freeport's Grasberg gold and copper mine, located in West Papua, Indonesia, was constructed between 1967 and 1972, and has been operational ever since. The original mine concession included the customary lands of two indigenous tribal groups, the Amungme and the Kamoro. Over time, as both the mine and the region have developed, employment opportunities and better living conditions have attracted other indigenous tribes (including the Dani, Ekari, Moni, Nduga and Damal) to the area. As their numbers increased, the immigrant tribes became more established as strong, politically powerful stakeholder groups, pressing their claims to be recognized as indigenous groups entitled to compensation by the company for adverse impacts associated with the mining operation.

Source: IFC (2009)

Several challenges arise from in-migration:

- Increased pressure on and degradation of natural resources, particularly for projects established in remote, previously undisturbed locations.
- Rapid population growth and large increases in the numbers of people living within the project area can strain public infrastructure, services, and utilities.
- Migrants may undermine the well-being of some members of the original resident population by threatening their way of life and the basis of existing livelihoods.

Project-induced in-migration may substantially change the context in which a project operates. An influx of migrants may affect host communities by changing how they secure their livelihoods or by introducing other (potentially disruptive and undesirable) socio-economic and environmental changes. Whatever the nature of these impacts, they can increase project costs and risks, and ultimately may affect the business licence to operate.

6.5.4 Governance issues

Companies often engage in cross-sector or multi-stakeholder collaborations to facilitate their sustainable development programmes. Such collaboration can strengthen or supplement but should not replace the role of government.

The roles and responsibilities of governments and companies are sometimes blurred, as both may appear to be working toward similar objectives. Many corporate sustainability initiatives, whether focused on biodiversity, economic development or both, aim to help local communities improve their use of available resources in order to meet their needs. The same is true of many public development interventions.

In addition, and notwithstanding the potential for companies to assist local partners in project development and implementation, businesses are sometimes held accountable for delivering (or failing to deliver) outcomes that may lie far beyond their mandate, competence or expertise (IPIECA 2006). For example, companies can make positive contributions to human rights, but they also face both legal and practical limitations as a non-government actor. While businesses can promote respect for human rights by assessing and addressing their own impacts, their efforts cannot and should not replace the role of governments (for more detailed discussion see Ruggie 2010).

6.6 Enablers and recommendations

Business can address the dual challenges of social development and biodiversity conservation through a range of measures, including:

1. Advocate for alignment between government programmes: stronger linkages and better coordination between government policies and programmes that address ecosystem management and economic development would make it easier for business to align its own efforts.
2. Cooperate with other stakeholders operating in the same region and/or country: business can engage with government, donor agencies, and other industries in the areas where it operates. Partnerships enable companies to contribute to sustainable development while reducing the risk of being held responsible for delivering results that exceed their expertise. Multi-stakeholder partnerships can also enhance innovation and help to scale up pilot initiatives to benefit a wider region.
3. Conduct thorough 'upfront' work during social impact assessment in order to build more inclusive business models: meaningful social investment and CSR programmes should begin with a thorough socio-economic baseline. Where possible, firms should engage with community members in their homes, talk to people about how they earn their income, whether their children are in school, the type of health problems they have and how many people they support. This type of analysis is not typically part of a social impact assessment (SIA). There are new business-focused impact assessment tools which can support better decision making in this context (see Box 6.11).
4. Direct interventions in biodiversity and ecosystem services: conservation efforts are generally more welcome and effective if directed at improving the management and delivery of ecosystem services that local communities depend on, and at the direct and indirect drivers that are undermining these services. This starts with efforts to account explicitly for ecosystem services and to identify those services which are most relevant for poverty reduction.

Box 6.11 WBCSD Measuring Impact Framework

The complex relationships between environment and development create both risks and opportunities for businesses. Companies need a measurement methodology that can systematically capture the interactions between the full spectrum of business activities and host societies in order to better understand and manage these risks and opportunities. In 2008 the WBCSD launched the Measuring Impact Framework in response to requests by companies for a common approach to this challenge. The framework is the result of two years of collaboration with over 20 companies and 15 external stakeholders from academia, NGOs and government.

The Measuring Impact Framework aims to help companies measure and assess their contribution to development and use this understanding to have more informed conversations with stakeholders, and ultimately make better investment and operational decisions. As such, it was not designed as a reporting or benchmarking tool, but rather as a methodology which can be adapted for a company, sector, country, specific project or operation. The methodology has four basic steps:

1 setting boundaries;
2 measuring direct and indirect impacts;
3 assessing contribution to development;
4 prioritizing management response.

The methodology complements and enhances traditional tools and approaches for environmental, social and health impact assessment (ESHIA). Indicators and data on interactions with BES will often be important components of the measurement of direct and indirect impacts, in Step 2. This data along with additional information on other aspects of a company's activities will allow a more holistic assessment of the firm's contribution to development (Step 3) and also help develop an informed prioritization of management responses (Step 4).

Source: www.wbcsd.org

Acknowledgements

Kit Armstrong (Plexus Energy), Tim Buchanan (BSR), Toby Croucher (Repsol), Juan Gonzalez-Valero (Syngenta), Michael Oxman (BSR).

Notes

1 'Bottom of the pyramid' (or 'base of the pyramid') is a phrase used to describe individuals in the largest and poorest socio-economic group (Prahalad 2006).

2 The anti-malarial activities of Artemether and Lumefantrine target the food vacuole of the malaria parasite. When the parasites infect human red blood cells (erythrocytes) they ingest and degrade haemoglobin and concentrate the iron in the form of 'haem' in a food vacuole. Artemether and Lumefantrine are thought to interfere with the process of detoxification of haem to haemozoin, and thus disrupt the feeding process of the parasite.

3 Novartis Annual Reports from 2004 to 2009 give a sum value of US$981 million (see the Novartis access-to-medicine tables, to which the years 2001 and 2002, as well as the first quarter of 2010, have been added). The estimated value of US$1 billion represents the financial value if the programme had been a for-profit enterprise and is calculated using the total number of treatments shipped, multiplied by the ex-factory price of Coartem to private-sector purchasers in malaria-endemic developing countries, less payments to Novartis to cover costs under the terms of its partnership with the WHO.

References

Alcoa (2010) *Juruti – Sustainable Development in Amazonia.* URL: www.alcoa.com/brazil/en/custom_page/environment_juruti_fundo.asp (last accessed 24 October 2010).

al-Janabi, S. and Drews, A. (2010) Genetic Resources as 'Biodiversity value added' for their providers, their users and mankind, ABS Capacity Development Initiative for Africa (GTZ), contribution to TEEB based on material available at URL: www.novartis.com/newsroom/corporatepublications/index.shtml, and URL: www.corporatecitizenship.novartis.com/downloads/business-conduct/Biodiversity.pdf

Anglo American (2005) *Anglo Zimele: Creating Sustainable Businesses, an Independent Management Review*. URL: www.angloamerican.com/aal/siteware/docs/zimele_independentreview.pdf (last accessed 20 October 2010).
Anglo American and IFC (2008) *The Anglo Zimele Model: A Corporate Risk Capital Facility Experience*, IFC Business Linkages Practice Notes. URL: www.ifc.org/ifcext/advisoryservices.nsf/AttachmentsByTitle/Business_Linkages_+Practice_Notes_Zimele/$FILE/Zimele+Manual+FINAL.pdf (last accessed 20 October 2010).
Ash, N. and Jenkins, M. (2007) *Biodiversity and Poverty Reduction: The Importance of Biodiversity for Ecosystem Services*, UNEP-WCMC, Cambridge, UK.
Banks, G. (2009) 'Activities of TNCs in extractive industries in Asia and the Pacific: Implications for development', *Transnational Corporations* 18(1): 43–60.
BP (2010) *Tangguh LNG*. URL: www.bp.com/sectiongenericarticle.do?categoryId=9004779&contentId=7008759 (last accessed 10 November 2010).
Brundtland, G. (ed.) (1987) *Our Common Future: The World Commission on Environment and Development*, Oxford, Oxford University Press.
Cervantes-Godoy, D. and Dewbre, J. (2010) *Economic Importance of Agriculture for Poverty Reduction*, OECD Food, Agriculture and Fisheries Working Paper No. 23, OECD Publishing. doi: 10.1787/5kmmv9s20944-en.
Conservation International (CI) (2008) *New Loans for Coffee Farmers, Nature Reserves*. URL: www.conservation.org/FMG/Articles/Pages/loans_for_coffee.aspx (last accessed 21 April 2010).
Conservation International (CI) (n.d.) *Starbucks Ethical Coffee Sourcing – An Assessment Conducted by Conservation International*. URL: www.conservation.org/discover/partnership/corporate/Documents/CAFE_Practices_Executive_Summary.pdf (last accessed 21 October 2010).
Department for International Development (DFID) (2006) 'Eliminating world poverty – making governance work for the poor', White Paper on International Development presented to Parliament by the Secretary of State for International Development, HMSO, London.
Emergia Institute and Allianz Group (2010) *Learning to Insure the Poor: Microinsurance Report*, Allianz SE, Munich, Germany.
IFC (2004) *A Guide to Biodiversity for the Private Sector, Comsur: A Junior Mining Company's Efforts to Conserve Biodiversity in Bolivia*. URL: www.ifc.org/ifcext/enviro.nsf/AttachmentsByTitle/BiodivGuide_CaseStudy_Comsur/$FILE/Comsur.pdf (last accessed 11 June 2010).
IFC (2009) *Projects and People: A Handbook for Addressing Project-Induced In-Migration*, International Finance Corporation, Washington, DC.
IFC and BSR (2008) *Local Content in Supply Chain*. URL: www.commdev.org/section/_commdev_practice/local_content_in_supply_chain (last accessed 17 May 2010).
IPIECA (2006) *Partnerships in the Oil and Gas Industry*. URL: www.ipieca.org (last accessed 5 July 2010).
IPIECA and OGP (2002) *Key Questions in Managing Social Issues in Oil and Gas Projects*, Report No. 2.85/332. URL: www.ogp.org.uk/pubs/332.pdf
Jenkins, B. and Ishikawa, E. (2010) *Scaling Up Inclusive Business: Advancing the Knowledge and Action Agenda*, International Finance Corporation and the CSR Initiative at the John F. Kennedy School of Government, Washington, DC.
Johnstone, N. and Wood, L. (eds) (2001) *Private Firms and Public Water: Realising Social and Environmental Objectives in Developing Countries*, Edward Elgar, Cheltenham, UK.
MA (2005) *Ecosystems and Human Well-being: Opportunities and Challenges for Business and Industry*, Island Press, Washington, DC.
McNeely, J. and Scherr, S. (2002) *Ecoagriculture: Strategies to Feed the World and Save Wild Biodiversity*, Island Press, Washington, DC.
Macqueen, D. (2008) *Supporting Small Forest Enterprises: A Cross-Sectoral Review of Best Practice*, IIED Small and Medium Forestry Enterprise Series No. 23, IIED, London, UK.
Mellor, J. (2002) *Poverty Reduction and Biodiversity Conservation: The Complex Role for Intensifying Agriculture*, WWF Macroeconomics for Sustainable Development Program Office, Washington, DC.
Milanovic, B. (2006) 'Global Income Inequality: A Review', *World Economics* 7(1) (January–March): 131–57.
Nelson, J. (2006) *Leveraging the Development Impact of Business in the Fight Against Global Poverty*, Corporate Social Responsibility Initiative Working Paper Series No. 22, John F. Kennedy School of Government, Harvard University, Cambridge, MA.

ODI and EAP (2007) *Learning From AMEC's Oil and Gas Asset Support Operations in the Asia Pacific Region with Case Study of the Bayu-Undan Gas Recycle Project, Timor-Leste*, Overseas Development Institute and Engineers Against Poverty, London.

Olsen, N. and Anstee, S. (2011) *Estimating the Costs and Benefits of Conserving the Tsitongambarika Forest in Madagascar*, IUCN and Rio Tinto, Gland.

Pagiola, S., Arcenas, A. and Platais, G. (2005) 'Can payments for environmental services help reduce poverty? An exploration of the issues and the evidence to date from Latin America', *World Development* 33(2): 237–53.

Pearce, D. (2005) *Investing in Environmental Wealth for Poverty Reduction – Environment for the MDGs*, Prepared for the Poverty Environment Partnership, UNDP, New York.

Prahalad, C.K. (2006) *The Fortune at the Bottom of the Pyramid: Eradicating Poverty Through Profits*, Wharton School Publishing, Upper Saddle River, NJ.

Rajan, R.G. and Zingales, L. (2006) 'Making capitalism work for everyone', *World Economics* 7(1) (January–March): 1–10.

Rodríguez, J.P., Beard, T.D. Jr., Bennett, E.M., Cumming, G.S., Cork, S.J., Agard, J., et al. (2006) 'Trade-offs across space, time, and ecosystem services', *Ecology and Society* 11(1): 28. URL: www.ecologyandsociety.org/vol11/iss1/art28/

RSB (2009) *RSB Principles and Criteria for Sustainable Biofuel Production*. URL: www2.epfl.ch/webdav/site/cgse/shared/Biofuels/Version%20One/Version%201.0/09-11-17%20RSB%20PCs%20Version%201%20%28clean%29.pdf (last accessed 21 October 2010).

Ruggie, J. (2010) Report of the Special Representative of the Secretary-General on the issue of human rights and transnational corporations and other business enterprises, 9 April. URL: http://198.170.85.29/Ruggie-report-2010.pdf.

SCBD (2002) *Bonn Guidelines on Access to Genetic Resources and Fair and Equitable Sharing of the Benefits Arising out of their Utilization*, SCBD, Montreal.

Shackleton, C., Shackleton, S., Gambiza, J., Nel, E., Rowntree, K. and Urquhart, P. (2008) *Links Between Ecosystem Services and Poverty Alleviation: Situation Analysis for Arid and Semi-arid Lands in Southern Africa*, Consortium on Ecosystems and Poverty in Sub-Saharan Africa.

Sievanen, L., Crawford, B., Pollnac, R. and Lowe, C. (2005) 'Weeding through assumptions of livelihood approaches in ICM: Seaweed farming in the Philippines and Indonesia', *Ocean and Coastal Management* 48(3–6): 297–313.

Starbucks (2010) *Responsibly Grown Coffee*. URL: www.starbucks.com/responsibility/sourcing/coffee (last accessed 21 October 2010).

TEEB in Policy Making (2011) *The Economics of Ecosystems and Biodiversity in National and International Policy Making* (ed. P. ten Brink), Earthscan, London.

UNDP (2004) *Unleashing Entrepreneurship: Making Business Work for the Poor*, United Nations Development Programme, New York, NY.

UNDP (2008) *Creating Value for All: Strategies for Doing Business with the Poor*, United Nations Development Programme, New York, NY.

UN Global Compact (2010) 'Overview of the UN Global Compact'. URL: www.unglobalcompact.org/AboutTheGC/ (last accessed 21 October 2010).

Unilever (2010a) *Sustainable Development Overview 2009: Creating a Better Future Every Day*. URL: www.unilever.com/images/sd_UnileverSDReport170310_amended_tcm13-212972.pdf (last accessed 21 October 2010).

Unilever (2010b) *Africa: Promoting Biodiversity and Alleviating Poverty in Ghana and Tanzania*. URL: www.unilever.com/sustainability/casestudies/economic-development/promoting-biodiversity-alleviating-poverty.aspx (last accessed 20 October 2010).

United Nations (2010) *The Millennium Development Goals Report 2010*, UN, New York, NY.

USAID (2010) *Alliance Industry Guide: Extractives Sector*, US Agency for International Development, Washington, DC.

Vodafone (2010) 'Access to communications in emerging markets'. URL: www.vodafone.com/start/responsibility/access_to_communications/emerging_markets/m-transactions.html (last accessed 21 October 2010).

WBCSD (2005) *Business for Development: Business Solutions in Support of the Millennium Development Goals*, WBCSD, Geneva.

WBCSD (2008) *Financial Services for the Unbanked: Vodafone*, WBCSD Case Study. URL: www.wbcsd.org/includes/getTarget.asp?type=DocDet&id=MjI5NTU (last accessed 21 October 2010).

Wells, S., Makoloweka, S. and Samoilys, M. (2007) *Putting Adaptive Management into Practice: Collaborative Coastal Management in Tanga, Northern Tanzania*, The World Conservation Union and Irish Aid, Nairobi, Kenya, p.197.

World Bank (2005a) *World Development Report 2005 – A Better Investment Climate for Everyone*, co-published by the World Bank and Oxford University Press, Oxford and New York, NY.

World Bank (2005b) *India's Water Economy: Bracing for a Turbulent Future*. URL: www.worldbank.org.in/WBSITE/EXTERNAL/COUNTRIES/SOUTHASIAEXT/INDIAEXTN/0,,contentMDK:20674796~pagePK:141137~piPK:141127~theSitePK:295584,00.html (last accessed 23 October 2010).

WRI (2005) *The Wealth of the Poor: Managing Ecosystems to Fight Poverty*, World Resources Institute, Washington, DC.

Chapter 7
Summary and Conclusion

Editor
Joshua Bishop (IUCN)

Contributing authors
Nicolas Bertrand (UNEP), Joshua Bishop (IUCN), William Evison (PwC), Sean Gilbert (GRI), Annelisa Grigg (Global Balance), Linda Hwang (BSR), Mikkel Kallesoe (WBCSD), Alexandra Vakrou (EC), Cornis van der Lugt (UNEP), Francis Vorhies (Earthmind)

Contents

Key messages

Biodiversity and ecosystem services are rising up the corporate agenda. Evidence provided in this book and in other publications, while still far too reliant on anecdotes and case studies, shows that more and more businesses see advantages in taking biodiversity and ecosystem services (BES) seriously.

Biodiversity loss and the decline in ecosystem services affect every business. Most environmental indicators point to continuing decline in natural capital, suggesting that economic development on the basis of 'business as usual' is unsustainable. Virtually all business sectors impact and depend on BES, creating both business risks and opportunities. The failure to recognize and manage BES impacts and dependencies can result in loss of market share, lower profit margins and missed opportunities for business innovation.

Biodiversity loss and ecosystem decline cannot be considered in isolation from other trends. The continuing loss of biodiversity and associated decline in ecosystem services is driven by growing and shifting markets, resource exploitation and climate change, among other factors. Equally, the decline of BES contributes to many of these other trends, implying the need for an integrated business response.

Effective integration of BES in business requires corporate governance reforms as well as better information systems. Trends in corporate BES targets, indicators and reporting practices suggest limited and inconsistent integration of BES across companies and sectors. A range of increasingly sophisticated environmental information systems have been developed and these can be readily expanded or extended to enable more consistent measurement, valuation, disclosure and reporting of BES impacts and dependence, both direct and indirect.

There is a growing toolbox, developed by, with and for businesses, to reduce or mitigate BES risk. Examples are available that show how these tools can enhance business value. Business action to reduce BES risk begins with the traditional environmental mitigation hierarchy (i.e. avoid, minimize, mitigate), although more and more companies are going further in an effort to achieve ecological 'neutrality' or better through restoration, offsets and biodiversity banking schemes.

New business models are emerging to create value by delivering BES benefits. 'Biodiversity business' models are emerging in established sectors as well as entirely new markets for biodiversity and ecosystem services. Business opportunities linked to BES are already attracting interest from investors and entrepreneurs. Public policy reforms can play a key role in stimulating private investment in BES opportunities.

Integrating BES in business can help improve social conditions. BES offers an opportunity to reinforce the contribution of business to sustainable livelihoods and poverty reduction. Many companies treat environmental and social responsibility as separate activities and therefore fail to realize potential synergies. Good examples are available showing how companies can integrate BES with social development.

Key knowledge and policy gaps must be addressed to accelerate integration of BES in business. Major gaps include the lack of agreed prices or values for many components of BES, inconsistent targets and reporting standards for business, excessive reliance on anecdotes, 'charity stories' and/or qualitative indicators to assess business BES performance, and, above all, the lack of enabling policy frameworks that can stimulate private investment in BES conservation and restoration as a profitable activity in its own right. These gaps not only limit business action but also make it harder to argue the case for conserving BES in business.

Business can take action immediately on BES:

- Identify the impacts and dependencies of your business on BES.
- Assess the business risks and opportunities associated with these impacts and dependencies.
- Develop BES information systems, set SMART (specific, measurable, achievable, relevant and time-bound) targets, measure and value performance, and report your results.
- Take action to avoid, minimize and mitigate BES risks, including in-kind compensation ('offsets') where feasible.
- Grasp emerging BES business opportunities, such as cost efficiencies, new products and new markets.
- Integrate business strategy and actions on BES with wider corporate social responsibility initiatives.
- Engage with business peers and stakeholders in government, NGOs and civil society to improve BES guidance and policy.

7.1 Summary of findings

This book has sought to make the case for integrating biodiversity and ecosystem services (BES) in business and enterprise, illustrated with examples from across sectors and regions. This chapter examines the implications of BES decline for business and other stakeholders, setting out an agenda for action informed by experience of what seems to work. Before doing so, however, we first review the main findings of the preceding chapters.

7.1.1 Awareness of biodiversity loss and ecosystem decline

Most authoritative indicators of the state of biological diversity ('biodiversity') show declines, indicators of pressures on biodiversity show increases and, despite some local successes and responses, the rate of biodiversity loss does not appear to be slowing (Butchart et al. 2010). Other recent assessments of ecological decline are equally disturbing (MA 2005a). The direct drivers of biodiversity loss include habitat loss and degradation, climate change, pollution, over-exploitation and the spread of invasive species (Baillie et al. 2004). Projections of the impacts of climate change, in particular, show continuing changes in the distribution and abundance of species and habitats, resulting in increasing species extinction (SCBD 2010).

Public awareness of biodiversity loss is growing, leading to changes in consumer preferences and purchasing decisions. Consumers appear to be more concerned about the environment today than just a few years ago (Taylor Nelson Sofres 2008). NGO campaigns, scientific research and media attention are part of the reason for this change but businesses are also showing leadership, indicated by the development and adoption of 'corporate responsibility' initiatives. As a result, more and more consumers favour ecologically certified goods and services (Box 7.1). This in turn increases pressure on business to review their value chains in order to ensure continued access to markets and protect against reputational risk. In some cases certification may be a requirement for market entry, while in others it may be a means to secure or increase market share (Bishop et al. 2008).

The financial services industry is also beginning to ask questions about BES. Investors are exploring new opportunities linked to BES but they are also increasingly

Box 7.1 Growth in markets for eco-certified products and services

- Global sales of organic food and drink amounted to US$46 billion in 2007, a threefold increase since 1999.
- US organic food sales alone accounted for 3.5 per cent of the nation's food market and increased by 15.8 per cent in 2008, more than triple the growth rate of the food sector as a whole in the same year.
- Sales of certified 'sustainable' forest products quadrupled between 2005 and 2007.
- Between April 2008 and March 2009, the global market for eco-labelled fish products grew by over 50 per cent, attaining a retail value of US$1.5 billion.

- In 2008–09, several brand owners and retailers added 'ecologically friendly' product attributes to their major consumer brands, often through independent certification schemes, including Mars (Rainforest Alliance cocoa), Cadbury (Fairtrade cocoa), Kraft (Rainforest Alliance Kenco coffee) and Unilever (Rainforest Alliance PG Tips).

Sources: FSC (2008); Marine Stewardship Council (2009); Organic Monitor (2009); Organic Trade Association (2009); Scott Thomas (2009)

concerned about potential risks (F&C Investments 2004; UNEP 2008). This is especially the case in the areas of project finance and re-insurance (Busenhart et al. 2007). Strategies employed include 'red-lining' investments in areas of high biodiversity, developing sector guidelines for environmentally sensitive sectors – for example, Rabobank has specific requirements regarding impacts on biodiversity for palm oil and soya – refraining from financing sectors in which a bank lacks knowledge of BES management and working with borrowers to improve their environmental performance and mitigate harm (Coulson 2009).

More generally, business is beginning to notice the threats posed by biodiversity loss and the decline in ecosystem services (WBCSD et al. 2006). Of over 1,200 global CEOs surveyed in 2009, 27 per cent expressed concern about the impacts of biodiversity loss on their business growth prospects (PwC 2010). Those expressing concern were more numerous in industries characterized by large direct impacts on biodiversity and in developing regions.

7.1.2 The value of biodiversity and ecosystem services

Environmentalists increasingly frame their analysis of biodiversity loss in terms of the benefits or 'ecosystem services' provided to people (MA 2005b). Ecosystem services enjoyed by people are economically significant and depend on both the diversity (quality) as well as the sheer amount (quantity) of genes, species and ecosystems found in nature (Table 7.1; see also: TEEB Foundations 2010; Chevassus-au-Louis et al. 2009; National Research Council 2005).

Table 7.1 Relationship between biodiversity, ecosystems and ecosystem services

Biodiversity	Ecosystem goods and services (examples)	Economic values (examples)
Ecosystems (variety and extent/area)	• Recreation • Water regulation • Carbon storage	Avoiding GHG emissions by conserving forests: US$3.7 trillion (net present value)
Species (diversity and abundance)	• Food, fibre, fuel • Design inspiration • Pollination	Contribution of insect pollinators to agricultural output: ~US$190 billion/year
Genes (variability and population)	• Medicinal discovery • Disease resistance • Adaptive capacity	25–50% of the US$640 billion pharmaceutical market is derived from genetic resources

Sources: Conserving forests: Eliasch (2008); insect pollination: Gallai et al. (2009); pharmaceuticals: TEEB in Policy Making (2011)

Scenario projections for the period 2000–2050 suggest improvement in most so-called 'provisioning' services – mainly food, fibre and other commodities – although this is likely to involve increased conversion of natural habitats and may come at the cost of further degradation in what the MA defined as 'supporting, regulating and cultural' services (MA 2005c). Some key provisioning services – notably fish and fresh water – may be at serious risk even in the short term. More generally, the continued rapid loss of biodiversity is expected to compromise future supplies of many ecosystem services and their associated economic outputs (Worm et al. 2006; Tilman et al. 2006; Gallai et al. 2009).

Biodiversity loss cannot be seen in isolation from other trends. The economic value of biodiversity and ecosystem services is a function of demand-side factors or underlying drivers of change (e.g. population growth and urbanization, economic growth, changing politics, preferences and environmental policy, developments in information and technology), as well as supply-side constraints (e.g. climate change, increasing scarcity of natural resources and/or declining quality of ecosystem services). Biodiversity loss and ecosystem decline are often closely linked to these and other major trends affecting business, as discussed in Chapter 1.

7.1.3 Measuring and managing business impacts and dependence

Far-sighted businesses can create opportunities from the greening of investor, client and consumer preferences. Businesses can influence consumer choice and behaviour by providing information about the sustainability of their products, as well as how to use and dispose of them responsibly. Companies can also develop 'smarter' products and services that help clients reduce their ecological footprint. The first step for business is to identify the impacts and dependencies of their products and services on biodiversity and ecosystem services, as discussed in Chapter 2.

All businesses benefit from biodiversity and ecosystem services, directly or indirectly; most businesses also have impacts on nature, positive or negative. Businesses that fail to assess their impacts and dependence on biodiversity and ecosystem services carry undefined risks and may neglect profitable opportunities.

A business commitment to manage BES begins with corporate governance and involves integrating biodiversity and/or ecosystem services into all aspects of business. Goals and targets for biodiversity and ecosystem services can be integrated into business risk and opportunity assessment, operations and supply chain management, and are increasingly being translated into financial accounting, audit and reporting systems. New and improved information systems are needed to support analysis and decision making about BES at corporate level, site/project level and product level, and for internal and external reporting of corporate performance, as discussed in Chapter 3.

Business can frame biodiversity and ecosystem targets in various ways – the challenge is to be SMART (specific, measurable, achievable, relevant and time-bound). Business efforts in relation to biodiversity and ecosystem services often start by identifying what to avoid (e.g. 'no-go' areas for exploration by extractive companies, prohibited technologies or sectors). Business can also express BES targets in more positive terms, such as 'reduce, reuse, recycle and restore', or adopt net balance approaches such as 'net positive impact' (see Chapter 4).

Measurement of biodiversity and ecosystem services is improving, but still challenging. Standard environmental performance indicators focus on direct inputs (e.g. water, energy or materials) and outputs (e.g. pollutant emissions, solid waste). Measurement of BES requires consideration of business impacts on all relevant components of biodiversity (i.e. genes, species, ecosystems), as well as the dependence of business operations on intangible biological processes (e.g. natural pest and disease control, nutrient cycles, decomposition). Life-cycle assessment (LCA) techniques and environmental management systems need to be expanded and refined to enable companies to assess BES along product life cycles and value chains (see Chapter 4 and Houdet et al. 2009). Despite such challenges, companies can begin to measure their impacts and dependence on biodiversity and ecosystem services using available metrics and reporting tools, even as they contribute to developing the field.

Economic valuation of biodiversity and ecosystem services can provide important information for business. Reliable methods are available to determine the economic values of BES (TEEB Foundations 2010), but more effort is needed to make ecosystem valuation methodologies accessible to corporate decision makers and to integrate the results into business decisions (WBCSD et al. 2011). The use of these methods in, for and by businesses can help strengthen the link from ecological impacts and dependence to the business bottom line.

Ultimately, the ability and interest of business to use such valuations in their financial accounts may depend on developments in accounting standards, financial disclosure requirements and environmental liability regulations. Financial regulators and accounting professionals are starting to provide guidance on how companies should report environmental issues – see for example US SEC (2010) and ICAEW and the UK Environment Agency (2009) – but more work is needed in partnership with other organizations with expertise on metrics and standards for biodiversity and for ecosystem services other than carbon emissions and mitigation. Many companies report their greenhouse gas emissions and mitigation efforts (CDP 2010). In contrast, biodiversity and ecosystem services are often treated superficially in company reports and are rarely seen as financially material in annual financial reporting. This may be due to insufficient clarity on reporting standards and the low priority assigned by reporting organizations, as well as weak regulatory guidance and incentives for business to address BES-related externalities. More generally, the lack of standard performance metrics for biodiversity and ecosystem services that can be used at company level and monitored continuously remains an obstacle to improved monitoring and disclosure. The Global Reporting Initiative (GRI) provides guidance and some basic indicators to start with, which can be refined to meet specific industry needs such as through GRI sector supplements (GRI n.d.).

7.1.4 Reducing biodiversity risk in business

Public acceptance of biodiversity loss is declining, leading to calls for low-impact production and, in some cases, monetary or in-kind compensation for impacts on biodiversity and ecosystems (e.g. from BBOP; see http://bbop.forest-trends.org). Many companies are exploring how to manage the adverse impacts of their activities on BES. A few companies have made public commitments to 'no net loss', 'ecological neutrality' or even 'net positive impact' on biodiversity, or on specific ecosystem services such as water resources, as discussed in Chapter 4. In some cases, even relatively simple ecological rehabilitation following resource extraction can deliver biodiversity benefits that may exceed those of the preceding land use.

Managing biodiversity risk involves looking beyond sites and products to the wider land and seascape. In the mining and oil and gas industries, for example, corporate environmental risk management has tended to focus on direct or primary impacts – those that result from site-level activities that could be avoided or mitigated through improved processes, procedures or technologies (Energy and Biodiversity Initiative 2003; ICMM 2006). However, increasing public scrutiny and more stringent regulations have led companies across a range of sectors to extend their risk horizon to include indirect or secondary impacts. This is echoed by growing interest in regional or landscape-level BES assessment and planning tools, product life cycle analysis and supply chain management, with more explicit and quantitative environmental indicators.

Effective biodiversity and ecosystem risk management may be facilitated by appropriate enabling frameworks and partnerships. These may include new markets for biodiversity-friendly products, investment screening processes that require attention to biodiversity impacts, and/or regulatory settings that pay close attention to biodiversity risks during the impact assessment process. Business risk management strategies also often involve public–private partnerships and stakeholder engagement (Tennyson with Harrison 2008).

7.1.5 Biodiversity as a business opportunity

Biodiversity and ecosystem services offer opportunities for all business sectors (Bishop et al. 2008). The integration of BES into business can create significant added value for companies, by ensuring the sustainability of supply chains, or by penetrating new markets and attracting new customers. Numerous examples are provided in Chapter 5.

Policies and procedures to manage biodiversity and ecosystem risk can also help identify new business opportunities, such as:

- reducing input costs through improved efficiency;
- developing and marketing low-impact technologies;
- managing and designing projects to reduce their footprint; and
- professional services in ecological risk assessment and management or adaptation (Hanson et al. 2008).

Biodiversity or ecosystem services can be the basis for new businesses. Conserving biodiversity and/or using it sustainably and equitably can be the basis for unique value propositions, enabling entrepreneurs and investors to develop and scale up 'biodiversity businesses', as discussed in Chapter 5.

The case for biodiversity as a business opportunity is perhaps most apparent in ecotourism, organic agriculture and sustainable forestry, where there is growing demand for 'sustainable' goods and services, as noted above. More generally, some estimates suggest that sustainability-related global business opportunities in natural resources (including energy, forestry, food and agriculture, water and metals) may be in the range of US$2–6 trillion by 2050, in 2008 prices (WBCSD 2010). If accurate, these projections suggest that the private sector will play an increasingly important role in natural resource management.

Tools for building biodiversity business are in place or under development. Critical market-based tools for capturing BES opportunities, such as biodiversity performance standards for investors, biodiversity-related certification, assessment and reporting schemes, and voluntary incentive measures, are available or under development and could be promoted across all business sectors and markets. One key tool is the International Finance Corporation's (IFC) Performance Standard 6 (PS6) on Biodiversity Conservation and Sustainable Management of Living Natural Resources (IFC 2011). This not only guides the investments of the IFC – the private sector arm of the World Bank – but also influences the investments of some 60 large, multinational banks that have adopted the Equator Principles, which call for adherence to IFC Performance Standards for project financing above US$10 million in emerging markets (Equator Principles n.d.).

Biodiversity and ecosystem service markets are slowly emerging, alongside markets for carbon. Effective responses to biodiversity loss and the decline in ecosystem services will require significant changes in economic regulations, incentives and markets (WBCSD and IUCN 2007). The global carbon market grew from virtually nothing in 2004 to over US$140 billion in 2009, largely as a result of new regulations driven by concern about climate change (Kossoy and Ambrosi 2010). New markets for biodiversity 'credits' and intangible ecosystem services such as watershed protection are also emerging, providing new environmental assets with both local and international trading opportunities (Madsen et al. 2010).

A first major market opportunity is likely to be reducing emissions from deforestation, forest degradation and related land-based carbon offsets, or REDD+ (TEEB in Policy Making 2011). Although designed mainly to address climate change, REDD+ is likely to deliver significant biodiversity benefits through the conservation of natural forests (Miles and Kapos 2008). Another potential market opportunity is the Green Development Initiative (GDI), a proposed innovative financial mechanism currently under discussion in the Convention on Biological Diversity and other fora (GDI 2011).

Appropriate public policies can create an enabling framework for new BES business. Inspired by the rapid development of global carbon markets and experience with markets for other ecosystem services (e.g. water markets in Australia, wetland mitigation banking in the USA), policy makers are experimenting with a range of regulatory reforms to support the development of environmental markets and business. Experience shows that the establishment of efficient ecosystem services markets requires several conditions to be met, involving inputs from financial and market experts as well as government (see Table 7.2). There is an opportunity for business to get involved in pilot schemes and help design efficient enabling conditions for such markets.

7.1.6 Synergies between business, biodiversity and development

Economic and social development generally involves more consumption and open markets, both highly correlated with business development but also often associated with biodiversity loss and ecosystem decline. The challenge is to reinforce economic development strategies that are ecologically sustainable, socially equitable and good for business, as discussed in Chapter 6.

Table 7.2 Enabling markets for biodiversity and ecosystem services

Financial	Regulatory	Market
• Adoption of integrated social, environmental and financial reporting • Clearly defined BES credits and debits • Insurability of BES assets • Investor awareness and support for commercial ventures • Competitive risk/reward profile • Combined ecosystem, business development and financial expertise	• Secure use and/or property rights over ecosystem assets and services • Clear baselines in order to assess the 'additionality' of BES investments • Approved standards and methods for assessing debits and credits • Fiscal incentives (e.g. tax credits for conservation) • Legal authority to trade ecosystem credits/debits (including internationally) • Adequate capacity to enforce regulations • Reform of corporate reporting and disclosure regulations to encourage integrated social, environmental and financial reporting, including BES	• Clearly defined asset classes • Efficient project approval processes • Modest transaction costs • Widely accepted monitoring, verification and enforcement systems • Linked registries to record transactions (especially for intangibles, e.g. offsets) • Competitive intermediary services (e.g. brokers, validators)

Source: PwC for TEEB

Good governance and clear property rights are essential for business development, environmental protection and poverty reduction. Better understanding of how different governance arrangements and especially property rights contribute to biodiversity loss and ecosystems degradation is needed in order to design responses that are not only ecologically sustainable but also socially acceptable. Reform of natural resource tenure, access rights and benefit-sharing arrangements can be an important complement to successful corporate community engagement.

There are potential synergies between business, conservation and poverty reduction, but these are not realized automatically. Biodiversity and ecosystem services are not routinely considered in most corporate decision making related to social investment. Some companies have programmes that support biodiversity conservation and separate initiatives that support local economic development. In many cases these initiatives are in conflict or fail to realize potential synergies, although a few companies have found ways to combine biodiversity and ecosystems with their social programmes, as highlighted in Chapter 6.

7.2 An agenda for action by business and other stakeholders

The business case for conserving biodiversity and ecosystem services is strong, and getting stronger. This book argues that companies that understand and manage the risks presented by biodiversity loss and ecosystem decline, establish operational models that are flexible and resilient to these pressures, and move quickly to seize BES-related business opportunities, are more likely to thrive in the future. Just as climate change has stimulated carbon markets and new business models, biodiversity and ecosystem services also offer opportunities for investors and entrepreneurs.

However, there is a need to agree priorities and adopt an agenda for action – by business leaders, accountancy and other professional bodies, regulators, government agencies and other stakeholders – otherwise significant change is unlikely.

7.2.1 Review of guidance for business on biodiversity and ecosystem services

In response to increasing awareness of and concern about biodiversity loss, there is a growing body of government policy, legislation and regulation, as well as numerous voluntary principles, declarations, charters, guidelines, handbooks, initiatives and tools. Some of these have been developed by businesses themselves, typically through industry associations, while other initiatives have come from government agencies or NGOs but are mainly targeted at business.[1] This book focuses more on voluntary initiatives rather than public policies and mandatory requirements, which are discussed in detail in a companion volume (TEEB in Policy Making 2011).

A range of general and sector-specific guidelines on biodiversity and/or ecosystems have been developed, while both industry and environmental organizations have set up various initiatives dedicated to engaging business in BES, environmental issues or sustainability more broadly. Annex 7.1 provides a comparison of several existing declarations, initiatives, guidelines and tools that aim to influence business practice in relation to biodiversity and/or ecosystem services. These are assessed in terms of the following aspects.

- **Scope** – sector(s), sub-national, national, regional, global.
- **Provenance** – from industry, NGO-led, government, multi-stakeholder.
- **Ecological focus** – biodiversity, ecosystems, specific services or combinations.
- **Guidance on metrics** – such as processes versus performance, monetary versus non-monetary indicators.
- **Risk management** – actions or methods to enhance biodiversity risk assessment and management.
- **Business opportunity** – actions to stimulate business investment in market opportunity based on biodiversity conservation or sustainable use.
- **Enabling frameworks** – relevant messages on enabling frameworks (e.g. user fees, environmental liability, disclosure obligations).
- **Poverty and social issues** – consideration of sustainable development, equity and poverty reduction.
- **Audit requirements** – internal and/or external verification commitments and procedures.

The proliferation of initiatives, such as those listed in Annex 7.1, shows that business is increasingly seen by many stakeholders as a crucial participant in efforts to halt biodiversity loss and ecosystem decline. More importantly, the number of initiatives that are led by industry, across a range of sectors, suggests that the business community itself increasingly sees a need to address the issue.

What is also clear is the growing sophistication of tools designed to help business minimize biodiversity risks in site selection or raw material sourcing, and to assess ecosystem service values and trends of relevance to business. There is widespread appreciation that businesses need clear and credible metrics to set realistic targets and to assess their performance in relation to biodiversity and ecosystem services, but also statements in many of the documents reviewed for Annex 7.1 that adequate BES metrics for business are still lacking (noted also in Chapter 3).

There appears to be less consistency across initiatives with respect to the need for and/or potential of regulatory reforms to ensure a level playing field and provide incentives for business action on biodiversity and ecosystems. Those initiatives that highlight this area typically call for 'market-based' or 'business-friendly' regulations, including tax credits and other government financial incentives, and less frequently for the introduction of markets for ecosystem services (e.g. 'biodiversity banking'). Similarly, the initiatives vary in their recognition of or enthusiasm for independent audit and assurance mechanisms, which can validate the claims of business and assess their performance over time.

These and other features of the existing guidance on business, biodiversity and ecosystems echo the findings of this book. In short, people working with and in business are increasingly aware of the benefits provided by biodiversity, the value of ecosystem services and the risks of biodiversity loss. This is accompanied by an emerging consensus around the need to strengthen business management systems and build supportive public policy frameworks that reflect more fully the ecological, economic and business values at stake. In addition, many initiatives:

- recognize the likelihood of increasingly stringent government regulations aimed at halting biodiversity loss and restoring ecosystem services;
- promote more systematic, risk-based valuation, accounting and disclosure of biodiversity assets, liabilities and performance; and
- consider biodiversity and ecosystems as business opportunities as well as risks, including opportunities to reduce costs, gain market share or develop entirely new products and services.

On the other hand, most existing initiatives fail to acknowledge the links between biodiversity conservation, ecosystem management and poverty reduction, as noted also in Chapter 6. More generally, it is difficult to assess the effectiveness of most existing business, biodiversity and ecosystem initiatives, such as those reviewed here and listed in Annex 7.1, partly because many are recent and do not have a long track record, but also because most of them do not publish independent evaluation reports, as noted above.

7.2.2 TEEB recommendations for business and other stakeholders

Based on the findings set out in the preceding chapters, together with the results of the brief review of existing business and biodiversity initiatives described above, we can identify several areas where further work is needed, as well as a number of concrete steps that companies can take immediately to integrate biodiversity and ecosystem services in their strategy and operations.

First, the accounting profession and reporting standards bodies should accelerate efforts, in partnership with others, to provide protocols and metrics for improved disclosure and audit/assurance of BES impacts and actions. Both general and sector-specific guidance is available for business on how to identify and address both the risks and opportunities associated with biodiversity and ecosystems. Governments, NGOs and business, often working together, have developed various principles, guidelines, handbooks and tools to help business address BES challenges. These initiatives often acknowledge the need for better metrics, including valuation, and sometimes call for enabling policy, including market-based incentives. Most existing initiatives are weak, however, at quantifying biodiversity impacts (the so-called 'externalities' of business) in terms of human welfare. Better methodologies for sector- and business-level quantification of biodiversity and ecosystem services values are needed, accompanied by appropriate reporting systems and stakeholder engagement to agree on their materiality. Credible audit and assurance mechanisms are needed to validate performance and the quality of disclosure, not only for business itself but also for many of the multi-stakeholder, business and biodiversity initiatives reviewed in this book.

Second, governments have an essential role to play in providing an efficient enabling and fiscal environment for biodiversity-friendly business and for setting clear and consistent limits on unacceptable practice. This includes removing environmentally harmful subsidies, offering tax credits or other incentives for conservation investments, establishing stronger environmental liability (e.g. performance bonds, offset requirements), developing new ecosystem property rights and trading schemes (e.g. water quality trading), supporting increased public access to relevant information through reporting and disclosure rules, and facilitating cross-sector collaboration (TEEB in Policy Making 2011).

Even without government action, based on existing science and methods, there is much that business can and should do in relation to biodiversity and ecosystem services:

1 **Identify the impacts and dependencies of your business on BES.** The first step is to assess business impacts and dependencies on biodiversity and ecosystems, including both direct and indirect linkages throughout the value chain, using existing tools while also helping to improve them. To reinforce such initiatives, business associations and others can help evaluate business experience of applying existing voluntary environmental tools and guidelines, identify their relative strengths and weaknesses, and support the development of new or improved tools that may be better adapted to business needs and capacities.
2 **Assess the business risks and opportunities associated with these impacts and dependencies.** Based on this assessment, companies can identify the business risks and opportunities associated with their impacts and dependencies on BES, and educate their employees, owners and customers. Economic valuation of BES impacts and dependencies can help clarify risks and opportunities. An urgent priority is to develop a representative and robust case study library of ecosystem valuations across industry sectors, covering all stages of the business value chain and decision-making processes.
3 **Develop BES information systems, set SMART targets, measure and value performance, and report your results.** Biodiversity and ecosystem strategies for business are likely to include improved corporate information systems, development of quantitative BES targets and performance indicators, and their integration into wider business risk and opportunity management processes. A key step for building trust with external stakeholders, while creating peer pressure within industry, is for business to measure and report their BES impacts, actions and outcomes against clear targets. To support such efforts, there is a need to forge

closer links between existing environmental disclosure initiatives and to encourage institutional investors to demand more consistent and quantitative disclosure by investee companies of biodiversity impacts and mitigation efforts.

4 **Take action to avoid, minimize and mitigate BES risks, including in-kind compensation ('offsets') where feasible.** BES targets may build on the concepts of 'no net loss', 'ecological neutrality' or 'net positive impact' and include support for biodiversity offsets where appropriate. Industry associations will continue to play a key role in developing and promoting robust and effective biodiversity performance standards and impact mitigation guidelines for their members. The ultimate aim is to stimulate increased business investment in biodiversity conservation and ecosystem rehabilitation, based on systematic and transparent assessment of the appropriate level of commitment, together with credible verification of conservation actions.
5 **Grasp emerging BES business opportunities, such as cost-efficiencies, new products and new markets.** There is growing interest in harnessing the profit motive to support sustainable natural resource management and conservation, inspired in part by the development of carbon trading and other environmental markets. Entrepreneurs and investors around the world have developed new business models that aim to generate biodiversity benefits as well as attractive financial returns. However, few biodiversity business ventures have attracted significant investment capital or achieved significant sales. Barriers to the growth of biodiversity business include conceptual, technical and political challenges. Business can support the growth of 'green' products and markets by working with other stakeholders to analyse and promote solutions that will attract private finance and achieve the scale required to make significant contributions to conservation.
6 **Integrate business strategy and actions on BES with wider corporate social responsibility initiatives.** There is potential to enhance both biodiversity status and human livelihoods, and help reduce global poverty, through the integration of BES in corporate sustainability and community engagement strategies. This requires closer coordination of business social and environmental strategies to identify and realize potential synergies while avoiding conflicts between BES and development initiatives.
7 **Engage with business peers and stakeholders in government, NGOs and civil society to improve BES guidance and policy.** Business can bring significant capacity to natural resource management and conservation efforts and has a major role in halting the decline in natural capital. However, in contrast to well-developed business positions on climate policy, there are few clear expressions of a business perspective on policies to address biodiversity loss and ecosystem decline. Business must participate actively in public policy discussions to advocate appropriate regulatory reforms, while also developing complementary voluntary guidelines. An immediate priority for business is to review the targets and indicators of the CBD Strategic Plan, agreed by most of the world's governments in 2010, so that companies can assess and demonstrate how their efforts contribute to wider commitments on biodiversity. This may include business engagement in the preparation or revision of National Biodiversity Strategies and Action Plans (NBSAPs), to ensure that new government targets and policy initiatives facilitate and encourage business contributions.

It is increasingly clear that biodiversity and ecosystem services have economic value. Both the conservation of biodiversity and the rehabilitation or maintenance of ecosystem services are directly relevant to business. For some businesses, this may involve mitigating biodiversity risks, while for others it may involve seizing new biodiversity opportunities. For all businesses, biodiversity loss and ecosystem decline are issues that can no longer be ignored. This review of the economics of ecosystems and biodiversity suggests that it is in the interest of all businesses to become 'biodiversity-positive'.

Acknowledgements

Edgar Andrukeitas (GIZ), Stefan Schaltegger (Leuphana University Lüneburg), Uwe Beständig (Leuphana University Lüneburg).

Note

1 One example of the growing literature promoting voluntary action by business on biodiversity and ecosystems is the 'Corporate Biodiversity Management Handbook' (Schaltegger and Beständig 2010). This document starts with a review of the drivers of biodiversity loss (i.e. habitat change, pollution, climate change, invasive alien species, over-exploitation) and then links these drivers to corporate structure (i.e. procurement, production, marketing, sales and distribution, research and development, human resources), as well as to business 'fields of action' (i.e. sites and facilities, supply chains, product, production, transport, personnel) and generic components of the 'business case' (i.e. cost, reputation and brand, sales and price, risk, innovation, the business model). In this way, the handbook helps translate the concepts and concerns of biodiversity conservation and ecosystem management into the general language and practice of business.

References

Baillie, J.E.M., Hilton-Taylor, C. and Stuart, S.N. (eds) (2004) *2004 IUCN Red List of Threatened Species: A Global Species Assessment*, IUCN, Gland, Switzerland, and Cambridge, UK. URL: http://data.iucn.org/dbtw-wpd/commande/downpdf.aspx?id=10588&url=www.iucn.org/dbtw-wpd/edocs/RL-2004-001.pdf (last access 23 June 2010).

Bishop, J., Kapila, S., Hicks, F., Mitchell, P. and Vorhies, F. (2008) *Building Biodiversity Business*, Shell International Limited and IUCN, London, UK, and Gland, Switzerland (March). URL: http://data.iucn.org/dbtw-wpd/edocs/2008-002.pdf (last access 23 June 2010).

Busenhart, J., Baumann, P., Orth, M., Schauer, C. and Wilke, B. (2007) *Insuring Environmental Damage in the European Union*, Technical Publishing, Casualty. SwissRe, Zurich.

Butchart, S. H. M. et al. (2010) *Global Biodiversity: Indicators of Recent Declines, Science* 328(5982): 1164–1168, 29 April, DOI: 10.1126/science.1187512.

CDP (2010) URL: www.cdproject.net/en-US/Pages/HomePage.aspx

Chevassus-au-Louis, B., Salles, J.-M., Bielsa, S., Richard, D., Martin, G., Pujol, J.-L. (2009) *Approche économique de la biodiversité et des services liés aux ecosystems: contribution à la décision publique*, Rapport du CAS, Paris.

Coulson, A. (2009) 'How should banks govern the environment? Challenging the construction of action versus veto', *Business Strategy and the Environment* 18(3):149–61 (May).

Eliasch, J. (2008) *Climate Change: Financing Global Forests, The Eliasch Review*, Earthscan, London. URL: www.occ.gov.uk/ activities/eliasch/Full_report_eliasch_review(1).pdf (last accessed 23 June 2010).

Energy and Biodiversity Initiative (2003) *Integrating Biodiversity into Oil and Gas Development*, EBI Report. URL: www.theebi.org/products.html (last accessed 23 June 2010).

Equator Principles (n.d.) 'Equator Principles'. URL: www.equator-principles.com/ (last accessed 23 June 2010).

F&C Investments (2004) *Is Biodiversity a Material Risk for Companies? An Assessment of the Exposure of FTSE Sectors to Biodiversity Risk* (September), originally published by ISIS Asset Management.

FSC (2008) *Facts and Figures on FSC Growth and Markets*. URL: www.fsc.org/fileadmin/web-data/public/document_center/powerpoints_graphs/facts_figures/Global-FSC-Certificates-2010-05-15-EN.pdf (last accessed 23 June 2010).

Gallai, N., Salles, J.-M., Settele, J. and Vaissière, B.E. (2009) 'Economic valuation of the vulnerability of world agriculture confronted with pollinator decline', *Ecological Economics* 68(3): 810–21.

GDI (2011) 'Conserving our planet, hectare by hectare'. URL: http://gdi.earthmind.net. See also: UNEP/CBD/WG-RI/3/INF/13, UNEP/CBD/COP/10/INF/15.

GRI (n.d.) *Sector Supplements*. URL: www.globalreporting.org/ReportingFramework/SectorSupplements/

Hanson, C., Ranganathan, J., Iceland, C. and Finisdore, J. (2008) *The Corporate Ecosystem Services Review: Guidelines for Identifying Business Risks and Opportunities Arising from Ecosystem Change*. WRI, WBCSD and Meridian Institute, Washington, DC. URL: http://pdf.wri.org/corporate_ecosystem_services_review.pdf (last accessed 28 June 2010).

Houdet, J., Pavageau, C., Trommetter, M. and Weber, J. (2009) *Accounting for Changes in Biodiversity and Ecosystem Services from a Business Perspective: Preliminary Guidelines towards a Biodiversity Accountability Framework*, Cahier no. 2009-44, Departement d'Economie, Ecole Polytechnique, Centre National de la Recherche Scientifique, Palaiseau (November). URL: http://hal.archives-ouvertes.fr/docs/00/43/44/50/PDF/2009-44.pdf (last accessed 28 June 2010).

ICAEW and the UK Environment Agency (2009) *Turning Questions into Answers: Environmental Issues and Annual Financial Reporting*, Sustainable Business Series, Institute of Chartered Accountants in England and Wales and the Environment Agency, London. URL: www.environment-agency.gov.uk/static/documents/Business/TECPLN8045_env_report_aw.pdf

IICMM (2006) *Good Practice Guidance for Mining and Biodiversity*, ICMM, London.

IFC (2011) *Performance Standard 6: Biodiversity Conservation and Sustainable Management of Living Natural Resources*, IFC, Washington, DC. URL: www.ifc.org/ifcext/sustainability.nsf/Content/EnvSocStandards (last accessed: 29 May 2011).

Kossoy, A. and Ambrosi, P. (2010) *State and Trends of the Carbon Market 2010*, World Bank, Washington, DC. (May) URL: http://siteresources.worldbank.org/INTCARBONFINANCE/Resources/State_and_Trends_of_the_Carbon_Market_2010_low_res.pdf (last accessed 28 June 2010).

MA (2005a) *Ecosystems and Human Well-being: Biodiversity Synthesis*, World Resources Institute, Island Press, Washington, DC. URL: www.millenniumassessment.org/documents/document.354.aspx.pdf (last accessed 23 June 2010).

MA (2005b) *Ecosystems and Human Well-being, Summary for Decision Makers*, Island Press, Washington, DC. URL: www.millenniumassessment.org/documents/document.356.aspx.pdf (last accessed 23 June 2010).

MA (2005c) *Scenarios Assessment*, Island Press, Washington, DC. URL: www.millenniumassessment.org/en/Scenarios.aspx (last accessed 23 June 2010).

Madsen, B., Carroll, N. and Moore Brands, K. (2010) *State of Biodiversity Markets*, Ecosystem Marketplace, Washington, DC. URL: www.ecosystemmarketplace.com/documents/acrobat/sbdmr.pdf

Marine Stewardship Council (2009) *Annual Report 2008/2009*. URL: www.msc.org/documents/msc-brochures/annualreport-archive/MSC-annual-report-2008-09.pdf/view?searchterm=annual%20report (last accessed 23 June 2010).

Miles, L. and Kapos, V. (2008) 'Reducing greenhouse gas emissions from deforestation and forest degradation: global land-use implications', *Science* 320, 1454–55. DOI: 10.1126/*science*.1155358

National Research Council (2005) *Valuing Ecosystem Services: Toward Better Environmental Decision-Making*, National Academies Press, Washington, DC.

Organic Monitor (2009) *Organic Monitor Gives 2009 Predictions*. URL: www.organicmonitor.com/r3001.htm (last accessed 28 June 2010).

Organic Trade Association (2009) *Organic Trade Association Releases Its 2009 Organic Industry Survey*. URL: www.npicenter.com/anm/templates/newsATemp.aspx?articleid=23917&zoneid=2 (last accessed 23 June 2010).

PwC (2010) *13th Annual Global CEO Survey 2010*. URL: www.pwc.com/gx/en/ceo-survey/index.html (last accessed: 15 June 2010).

SCBD (2010) *Global Biodiversity Outlook 3*. URL: www.cbd.int/doc/publications/gbo/gbo3-final-en.pdf (last accessed 23 June 2010).

Schaltegger, S. and Beständig, U. (2010) *Corporate Biodiversity Management Handbook: A Guide for Practical Implementation*, Federal Ministry for the Environment, Nature Conservation and Nuclear Safety (BMU), Berlin (June).

Scott Thomas, C. (2009) 'Organic foods are now "mainstream", says USDA', *Food and Drink Europe*, 14 September. URL: www.foodanddrinkeurope.com/Consumer-Trends/ Organic-foods-are-now-mainstream-says-USDA (last accessed 23 June 2010).

Taylor Nelson Sofres (2008) *Global Shades of Green*. URL: www.tns-us.com/greenlife/ (last accessed 23 June 2010).

TEEB Foundations (2010) *The Economics of Ecosystems and Biodiversity: Ecological and Economic Foundations* (ed. P. Kumar), Earthscan, London.

TEEB in Policy Making (2011) *The Economics of Ecosystems and Biodiversity in National and International Policy Making* (ed. P. ten Brink), Earthscan, London.

Tennyson, R., with Harrison, T. (2008) *Under the Spotlight: Building a Better Understanding of Global Business–NGO Partnerships*, International Business Leaders Forum, London, UK. URL: www.iblf.org/~/media/Files/Resources/Publications/Under_the_spotlight2008.ashx (last accessed 23 June 2010).

Tilman, D., Reich, P.B. and Knops, J.M.H. (2006) 'Biodiversity and ecosystem stability in a decade long grassland experiment', *Nature* 441 (1 June): 629–32, doi: 10.1038/nature04742

UNEP (2008) *Biodiversity and Ecosystem Services: Bloom or Bust? A document of the UNEP FI Biodiversity and Ecosystem Services Work Stream*, UNEP Finance Initiative, Geneva (March). URL: www.unepfi.org/fileadmin/documents/bloom_or_bust_report.pdf (last accessed 23 June 2010).

US SEC (2010) *Commission Guidance Regarding Disclosure Related to Climate Change*, Securities and Exchange Commission, 17 CFR Parts 211, 231 and 241, Release Nos. 33-9106; 34-61469; FR-82. URL: www.sec.gov/rules/interp/2010/33-9106.pdf

WBCSD (2010) *Vision 2050: The New Agenda for Business*. World Business Council for Sustainable Development, Geneva. URL: www.wbcsd.org/web/projects/BZrole/Vision2050-FullReport_Final.pdf (last accessed 23 June 2010).

WBCSD and IUCN (2007) *Markets for Ecosystem Services – New Challenges and Opportunities for Business and the Environment*, WBCSD and IUCN, Geneva and Gland. URL: www.wbcsd.org/DocRoot/7g8VZQpq0LeF1xNwsbGX/market4ecosystem-services.pdf (last accessed 28 June 2010).

WBCSD, IUCN, WRI and Earthwatch (2006) *Ecosystem Challenges and Business Implications*, World Business Council for Sustainable Development, Geneva.

WBCSD, ERM, IUCN and PwC (2011) *Guide to Corporate Ecosystem Valuation: A Framework for Improving Corporate Decision-Making*, World Business Council for Sustainable Development, Geneva (April). URL: www.wbcsd.org/web/cev.htm

Worm B., Barbier, E.B., Beaumont, N., Duffy, J.E., Folke, C., Halpern, B.S., et al. (2006) 'Impacts of biodiversity loss on ocean ecosystem services', *Science* 314 (3 November): 787–90.

Annex 7.1 Comparison of selected business, biodiversity and ecosystem declarations, initiatives, guidelines and tools

Annex 7.1 Comparison of selected business, biodiversity and ecosystem declarations, initiatives, guidelines and tools[1]

Name	Scope	Provenance	Ecological focus	Metrics	Risk management	New business opportunities	Enabling frameworks	Poverty and social	Audit
ARtificial Intelligence for Ecosystem Services (ARIES) Project[2]	Global; all sectors	Academia	Biodiversity and ecosystem services, with a focus on ecosystem services	Values ecosystem services for decision making	Provides a tool for management of risk	Provides a tool to protect opportunity	Provides a tool to identify/ assess opportunity	No specific reference	No specific requirement
Biodiversity in Good Company: Business and Biodiversity Initiative[3]	All sectors; initial focus mainly on German companies but operates internationally to provide services, e.g. consultancy, access to experts, workshops, roundtables	Government	Biodiversity in line with the CBD definition	Signatories commit to: develop biodiversity indicators; define realistic, measurable objectives that are monitored and adjusted every 2–3 years; publish achievements externally	Commit to analyse corporate activities with regard to their impacts on biological diversity	No specific reference	Voluntary initiative; companies commit to an internal review process and to disclose their activities, including presentation of activities to the Conference of the Parties of the CBD	Reference to fair and equitable sharing of benefits; no specific commitments to poverty reduction	Companies in the initiative are not subject to audit of their performance
Business and Biodiversity Offsets Programme[4]	Global; all sectors	NGO	Biodiversity and ecosystem services	Recommends a range of indicators to assess impacts and offset implementation. Quantification is stressed, with the aim of ensuring 'no net loss' of biodiversity from development projects	Strategic review of biodiversity-related risks is part of the offset design process	Strategic review of biodiversity-related opportunities is part of the offset design process	Integration of offsets in land-use planning frameworks; notes lack of a common currency for quantifying losses and gains, need to build capacity to design and mplement offsets; works with several governments to develop biodiversity offset and banking policy	Special attention to be paid to the rights of indigenous peoples and local communities	Developing assurance standards (principles, criteria and indicators) initially for voluntary adoption

Annex 7.1 (Cont'd)

Name	Scope	Provenance	Ecological focus	Metrics	Risk management	New business opportunities	Enabling frameworks	Poverty and social	Audit
Biodiversity Risk and Opportunity Assessment Tool[5]	Agricultural supply chain	NGO and corporate	Biodiversity and ecosystem services	Indicators developed for each identified high and medium risk and opportunity. Mitigation hierarchy used to develop action plan	Focus on identifying risks linked to impacts on biodiversity and ecosystem services	Offers guidance to identify and develop opportunities linked to biodiversity	No specific reference	Need to identify and allow for community reliance on natural resources	No specific reference
Canadian Business and Biodiversity Initiative[6]	Canada; all sectors	Government and industry	Biodiversity	No specific reference; intend to develop an award scheme	No specific reference	No specific reference	No specific reference	No specific reference	No specific reference
Convention on Biological Diversity: Jakarta Charter on Business and Biodiversity[7]	Global; all sectors	Convention on Biological Diversity, 3rd Business and Biodiversity Challenge Meeting (Jakarta, Indonesia, Nov.–Dec. 2009)	Biodiversity and ecosystem services	No net loss and net positive impact are both highlighted as a potentially useful framework; recognizes the need to improve quantity, quality and availability of data for decision making	Recognizes the need to incorporate biodiversity risk into business practices and policies	Recognizes the need to incorporate biodiversity opportunity into business practices and policies	Highlights need to: recognize value of biodiversity and ecosystem services in economic models and policies; combine voluntary activity with market mechanisms; promote eco-certification and encourage states to recognize biodiversity in procurement policies;	No specific reference	No specific reference

Annex 7.1 (Cont'd)

Name	Scope	Provenance	Ecological focus	Metrics	Risk management	New business opportunities	Enabling frameworks	Poverty and social	Audit
							develop multi-sector enabling framework on business and biodiversity; strengthen capacity; and create a policy environment that encourages business engagement		
The Corporate Biodiversity Management Handbook[8]	Global; all sectors	Government	Biodiversity in line with the CBD definition, including ecosystem services	Recommends use of indicators, provides examples but does not provide detail or recommendations. Examples are of non-monetary indicators	Risk mitigation is a feature throughout the handbook	Refers to opportunities linked to biodiversity and ecosystem services and nature as a source of innovation	No specific reference	No specific reference	Recommends that companies have a management system in place that includes periodic audit
Declaration of Biodiversity by Nippon Keidanren[9]	Japanese, but with international reach; all sectors	Industry association	Biodiversity and ecosystem services	No specific reference	No specific reference, but commit to improve identification and analysis of biodiversity impacts, to improve business operations on biodiversity, and to pursue operational activities with low impact on biodiversity	Commit to learning from nature, society's wisdom and traditions regarding nature, and to pursue management innovations by promoting the development of environmental technology	Commit to consider implementation of trading or offsetting measures based on an economic evaluation and to spearhead activities to build a society that will nurture biodiversity	Requirement to consider impact on local communities	No specific reference

Annex 7.1 (Cont'd)

Name	Scope	Provenance	Ecological focus	Metrics	Risk management	New business opportunities	Enabling frameworks	Poverty and social	Audit
Handbook for Corporate Action[10]	Global; all sectors	NGOs and industry association	Biodiversity and ecosystem services	Companies should monitor performance, but notes metrics are lacking	Sets out business case from a risk management perspective	Sets out business case from an opportunity realization perspective	No specific reference	Benefit sharing and stakeholder consultation recommended	Evaluation and review recommended
Equator Principles[11]	Global; for finance sector (project finance and advisory) but provides guidance that covers all sectors	Industry (and IFC)	Biodiversity and ecosystem services	No metrics set out in standard but accompanying guidance sets out requirement to have both process- and performance-based measures for site-level biodiversity action plan implementation, with a requirement for quantification where possible	Assess all impacts on biodiversity bearing in mind differing stakeholder values	Opportunities to enhance biodiversity should be considered but no specific reference to biodiversity business opportunities	Restoration, offset and compensation set out as mechanisms to achieve no net loss of biodiversity, where residual damage is unavoidable	Requires consideration of affected communities; recognizes need to understand social and cultural values and for prior informed consent	Commit to certification where possible – should be independent, cost-effective, stakeholder inclusive and transparent Equator Principle implementation itself is subject to third-party review on a project-by-project basis
European Business and Biodiversity Campaign[12]	European; all sectors	Government, NGO led	Biodiversity	No specific reference	No specific reference	Exchange information and disseminate tools to enable opportunity identification	Exchange information and tools as enabler of action	No specific reference	No specific reference
European Business and Biodiversity Platform[13]	European; initial focus on agriculture, food, forestry, extractive industry, financial services and tourism	Government (EU)	Biodiversity and ecosystem services	Developing a benchmarking and award system to recognize strong performance, metrics are not stipulated	Knowledge platform to provide examples of good practice and facilitate exchange of information; no specific reference to biodiversity risk management	Provides examples of good practice and exchange of information; no specific reference to opportunity realization	Enabling activities including information sharing, provision of tools and guidance. No specific reference to policy reform	No specific reference	No specific reference

Annex 7.1 (Cont'd)

Name	Scope	Provenance	Ecological focus	Metrics	Risk management	New business opportunities	Enabling frameworks	Poverty and social	Audit
Forest Footprint Disclosure Project[14]	Global; various sectors	NGO with finance sector support	Forest ecosystems	Requests information on metrics used to measure performance, e.g. volume of certified product purchased	Asks companies to disclose risks associated with impacts on forests	Asks companies to disclose opportunities associated with impacts on forests	No specific reference	No specific reference	Requests information on the use of third-party certification schemes
Biodiversity Accountability Framework[15]	Global; all sectors	NGO	Biodiversity and ecosystem services	Provides a qualitative accounting framework	Helps companies understand risks associated with their impacts and dependence on biodiversity and ecosystem services	Helps companies understand opportunities associated with their impacts and dependence on biodiversity and ecosystem services	Notes need to understand the relationship between resources and users; calls for new accounting framework that accounts for business and ecosystems, and associated accounting and fiscal instruments	No specific reference	Not an audit tool
Global Reporting Initiative[16]	Global; all sectors, with supplements for certain sectors, e.g. mining and metals, finance, oil and gas, food processing, construction, utilities	NGO with multi-stakeholder input	Biodiversity primarily, although some coverage of water-related ecosystem services	Quantitative indicators for biodiversity, mainly process measures	Requires disclosure of key risks for all 'material' sustainability issues and a process for addressing them, including biodiversity	Requires disclosure of key opportunities and a process for addressing them, including biodiversity	Voluntary only; no specific reference to enabling frameworks	Recognizes poverty issues, but no explicit link to biodiversity indicators	Requires disclosure of whether external or internal assurance is in place but not required
Integrated Biodiversity Assessment Tool[17]	Global; all sectors	NGO	Biodiversity	Does not specify metrics; provides data on location of sensitive sites/ protected areas	Provides a tool to identify potential risks	Provides a tool to identify potential opportunities	No specific reference	No specific requirement	No specific requirement

Annex 7.1 (Cont'd)

Name	Scope	Provenance	Ecological focus	Metrics	Risk management	New business opportunities	Enabling frameworks	Poverty and social	Audit
Integrated Valuation of Ecosystem Services and Tradeoffs (InVEST)[18]	Ambition is global but initially focused on a few regions; all sectors	NGO and academia	Biodiversity and ecosystems, with emphasis on ecosystem services	Monetary valuation of ecosystem services	Provides information for decision making and risk evaluation	Provides information for decision making opportunity evaluation	No specific reference	No specific requirement	No specific requirement
ICMM Good Practice Guidance[19]	Global; mining	Industry association	Biodiversity and ecosystem services	Advises the use of indicators to monitor success, refers to the GRI indicators	Recommends evaluation of risk using stakeholder input	Focus on opportunities for more conservation but not *business* opportunities	No specific reference	Stakeholder consultation is a key theme, but it is not designed to address poverty	Monitoring and evaluation recommended but not required
ICMM Sustainable Development Framework[20]	Global; mining and metals	Industry (ICMM members) based on mining, mineral and sustainable development consultation process	Biodiversity in the original principles, but GRI mining and mineral sectors supplement has introduced language on ecosystem services[21]	Commitment to report against the GRI (see below); some biodiversity and ecosystem services indicators; mainly qualitative indicators for ecosystem services	Contribute to the conservation of biodiversity and adopt integrated approaches to land-use planning	No specific reference	Requires evaluating impact and risk, disclosure of activity; working with IUCN to address issues relating to protected areas categories[22]	Commit to contribute to social and economic development of communities in which they operate	Audit framework developed for principles. Must be implemented by all members by Dec. 2009 or March 2010[23]

Annex 7.1 (Cont'd)

Name	Scope	Provenance	Ecological focus	Metrics	Risk management	New business opportunities	Enabling frameworks	Poverty and social	Audit
							identifies needs to work with other stakeholders to develop decision making processes and assessment tools that better integrate biodiversity conservation, protected areas and mining into land-use planning and management strategies, including 'no-go' areas		
IFC Performance Standard 6: Biodiversity Conservation and Sustainable Management of Living Natural Resources[24]	Global; all sectors	Industry (IFC), developed through stakeholder consultation	Biodiversity, with increased emphasis on ecosystem services in Version 2 (2011)	Calls for 'no net loss of biodiversity where feasible' in natural habitat and 'net positive gain' in 'critical habitat' but does not specify metrics; guidance sets out requirement to use both process- and performance-based measures for site-level biodiversity action plans, including quantification where possible	Assess all impacts on biodiversity bearing in mind differing stakeholder values	Opportunities to enhance biodiversity should be considered; no specific reference to biodiversity business opportunities	Restoration, offset and compensation set out as mechanisms to achieve no net loss, where residual damage is unavoidable	Requires consultation with affected communities and also 'free, prior and informed consent' for projects with potential significant adverse impacts on indigenous peoples	Commit to certification 'when available, for the living natural resource and country concerned'

Annex 7.1 (Cont'd)

Name	Scope	Provenance	Ecological focus	Metrics	Risk management	New business opportunities	Enabling frameworks	Poverty and social	Audit
IPIECA[25]	Global; oil and gas	Industry association	Biodiversity initially; later documents also cover ecosystem services	Use encouraged but no specific metrics are suggested	Biodiversity and ecosystem services management outlined as a risk management strategy	Opportunities for partnering and earning credits for conservation	No specific reference	Engagement with communities and indigenous peoples	Monitoring and review recommended
LIFE Certification[26]	Brazil; all sectors	NGO and industry	Biodiversity and ecosystem services	Companies receiving certification must have indicators in place, but not stated whether indicators must be quantitative	Focus on evaluation and management of impacts rather than business risk	Focus on evaluation and management of impacts rather than business opportunities	No specific reference	Benefit sharing is included in the scheme	Certification scheme involves external audit
Millennium Ecosystem Assessment: Business Synthesis[27]	Global; all sectors	Multi-stakeholder	Ecosystem services	Recommends identifying appropriate indicators	Sets out business risks associated with decline in ecosystem services	Describes business opportunities linked to the decline in ecosystem services	Calls for: increased policy integration; coordination among multilateral agreements; integration of ecosystem management goals within sectors and broader development planning frameworks; increased;	Recognizes the need to empower indigenous peoples	Encourages use of third- party certification as means to build and maintain trust among consumers/ customers

Annex 7.1 (Cont'd)

Name	Scope	Provenance	Ecological focus	Metrics	Risk management	New business opportunities	Enabling frameworks	Poverty and social	Audit
							transparency and accountability of government and the private sector enhanced capacity; better communication; empowerment of resource-dependent communities; resource management policies to take account of ecosystem services; elimination of harmful subsidies; development of new technologies; promotion of sustainable intensification in agriculture; increased use of economic instruments and market-based approaches		

Annex 7.1 (Cont'd)

Name	Scope	Provenance	Ecological focus	Metrics	Risk management	New business opportunities	Enabling frameworks	Poverty and social	Audit
Natural Value Initiative[28]	Global, food and beverage sectors, tobacco, pharmaceuticals, oil and gas, mining	Multi-stakeholder	Biodiversity and ecosystem services	Ecosystem Services Benchmark uses process-based metrics to generate quantified scores of corporate performance; exploring biodiversity, ecosystem services and financial materiality	Explores business risk in relation to a company's dependency and/or impact on ecosystem services	Highlights potential for market differentiation and access to new revenue streams	Integration of the five key criteria of the Ecosystem Services Benchmark in a company's business operations	No specific reference	Additional 'credit' awarded on evaluation if companies use internal and external audit processes
Conservation by Design[29]	Global; all sectors but aimed at conservation practitioners	NGO	Biodiversity and ecosystem services	Provides process and example of metrics for conservation, but not tailored to business	Risk management (from perspective of conservation management rather than business risk)	No specific reference	No specific reference; such measures would be one output of applying 'conservation by design'	No specific reference to poverty but emphasizes the need to work with communities	No specific reference
Sustainable Forest Finance Toolkit[30]	Global; investors focused on the forest sector	Industry-led (WBCSD Sustainable Forest Products Working Group)	Forestry, sustainable forest management, forest financing	No specific reference	Key resources: 1 Client and transaction screening tools 2 Briefing notes on key risks in the forestry sector	Briefing notes on opportunities related to sustainable forest management	No specific reference, but provides country briefings on drivers of risk and opportunity	Reference to human rights, stakeholder engagement, community well-being	Recommends that banks undertake internal audit of policy compliance. Clients are required to have their forests certified

Annex 7.1 (Cont'd)

Name	Scope	Provenance	Ecological focus	Metrics	Risk management	New business opportunities	Enabling frameworks	Poverty and social	Audit
					3 Portfolio management guidance 4 Principles for the development of a forestry policy 5 Guidance on the development of a procurement policy for forestry products				
Proteus[31]	Global; all sectors	NGO	Biodiversity	Does not specify metrics; provides data on location of sensitive sites/ protected areas	Provides a tool for risk management	Provides a tool to identify opportunity	No specific reference	No specific requirement	No specific requirement
UNEP Finance Initiative: Biodiversity and Ecosystem Services Work Stream (BESWS)	Global, financial institutions	UNEP and financial institutions	Ecosystem services and biodiversity	Use of metrics encouraged, such as the Natural Value Initiative's Ecosystem Services Benchmark (see above)	Focus on building the business case for biodiversity and ecosystem services from a financial sector perspective	Focus on the development of environmental markets, such as carbon credits related to REDD/ avoided deforestation	No specific reference	No specific reference	No specific reference
TWT Biodiversity Benchmark[32]	UK focused; all sectors	NGO	Biodiversity	Specifies criteria that a company must meet to be certified against the Benchmark; no specific requirement for companies to use metrics	Focus on evaluating and managing impacts rather than business risk	Focus on evaluating and managing impacts rather than business opportunities	No specific reference	No specific reference	Certification scheme, requires companies to have an internal audit process

Annex 7.1 (Cont'd)

Name	Scope	Provenance	Ecological focus	Metrics	Risk management	New business opportunities	Enabling frameworks	Poverty and social	Audit
Corporate Ecosystem Services Review[33]	Global; all sectors	NGO with industry association	Ecosystem services	No specific metrics required, focus on strategy development	Guides companies to assess risks associated with their impacts and dependence on ecosystem services	Guides companies to assess opportunities associated with their impacts and dependence on ecosystem services	Suggests a range of response strategies including policy reform and stakeholder collaboration	Requires engagement with various stakeholders including indigenous peoples and communities	No specific reference
Corporate Ecosystem Valuation[34]	Global; all sectors	NGOs, business and industry association	Ecosystem services	Provides guidance on monetary valuation of the business risks and opportunities associated with ecosystem change	Promotes ecosystem valuation as a tool to inform and help companies develop risk management strategies	Promotes ecosystem valuation as a tool to inform and help companies develop opportunity management strategies	Builds on the Corporate Ecosystem Services Review (see above); suggests that ecosystem valuation can help 'inform government regulations and policies' but does not call for specific policy reforms	No specific reference	No specific reference
Global Water Tool[35]	Global; all sectors	Industry association	Water	Develops GRI Water Indicators e.g. number of sites, employees, suppliers in water-stressed regions	Water-related risk management tool	Tool can be used to explore water-related implications of future operations	No specific reference	No specific reference	No specific requirement, although endorses GRI as a reporting framework

Annex 7.1 (Cont'd)

Name	Scope	Provenance	Ecological focus	Metrics	Risk management	New business opportunities	Enabling frameworks	Poverty and social	Audit
WEF Global Agenda Council on Ecosystems & Biodiversity[36]	Global; all sectors	Multi-stakeholder	Ecosystem services and biodiversity	No specific metrics provided, although calls on companies to 'measure and report your direct and indirect impacts on biodiversity and ecosystems, across the value chain', as a means to become 'biodiversity positive'	Examples of business risks from biodiversity loss and ecosystem degradation given for agriculture, construction and real estate, financial services, pharmaceuticals and personal care, and mining sectors	Examples of ecosystem service related opportunities given for agriculture, construction and real estate, financial services, pharmaceuticals and personal care, and mining sectors	Future development needs to be based on a minimum standard of no net loss of natural capital; value needs to be attributed to natural capital; carbon finance is identified as a lever to secure the values of other ecosystem services Refers to environmental governance, participation in decision making, equitable distribution of rights/benefits and enhancing awareness among public and government officials	Recognizes the poor as stewards of biodiversity	Certification is suggested as a means to verify products on environmentally sensitive markets

[1] The table is not intended to be exhaustive; a more comprehensive list of relevant initiatives is given in UNEP (2010) *Are You a Green Leader?* (www.uneptie.org/scp/business/publications/pdf/Are_you_a_green_leader_final_publication.pdf).

[2] ARtificial Intelligence for Ecosystem Services (ARIES) Project (www.ariesonline.org).

3 BMU and GTZ (2008) *Leadership Declaration for the Implementation of the UN Convention on Biological Diversity, An Initiative of the Federal Ministry for Environment, Nature Conservation and Nuclear Safety and Leading Companies* (www.business-and-biodiversity.de/en/homepage.html).

4 Business and Biodiversity Offsets Programme (http://bbop.forest-trends.org/guidelines/index.php).

5 British American Tobacco Biodiversity Partnership (www.batbiodiversity.org).

6 Canadian Business and Biodiversity Program (www.businessbiodiversity.ca/index.cfm).

7 CBD (2009) *The Jakarta Charter on Business and Biodiversity* (www.cbd.int/doc/business/jakarta-charter-busissness-en.pdf).

8 Schaltegger, S., and Beständig, U. (2010) *Corporate Biodiversity Management Handbook. A Guide for Practical Implementation.* Leuphana University, Lüneburg.

9 Nippon Keidanren (Japan Business Federation) (2009) *Declaration of Biodiversity* (www.keidanren.or.jp/english/policy/2009/026.html).

10 Earthwatch, IUCN and WBCSD (2002) *Business and Biodiversity: Handbook for Corporate Action* (www.wbcsd.org/web/publications/business_biodiversity2002.pdf).

11 The Equator Principles (www.equator-principles.com/).

12 European Business and Biodiversity Campaign (www.globalnature.org/30707/campaigns/eu-business-biodiversity-campaign/02_vorlage.asp).

13 European Business and Biodiversity Initiative (ec.europa.eu/environment/biodiversity/business/index_en.html).

14 Forest Footprint Disclosure Project (2009) *Forest Footprint Disclosure Request* (www.forestdisclosure.com).

15 Houdet, J. (ed.) (2010) *Integrating Biodiversity into Business Strategies: The Biodiversity Accountability Framework.* FRB and Orée, Paris.

16 Global Reporting Initiative (2006) *G3 Guidelines* (www.globalreporting.org/ReportingFramework/G3Guidelines/).

17 Integrated Biodiversity Assessment Tool (www.ibatforbusiness.org/).

18 Integrated Valuation of Ecosystem Services and Tradeoffs (InVEST) (www.naturalcapitalproject.org/InVEST.html).

19 ICMM (2006) *Good Practice Guidance for Mining and Biodiversity* (www.icmm.com/document/13).

20 International Council of Mining and Metals (2008) *Sustainable Development Framework* (www.icmm.com/our-work/sustainable-development-framework).

21 Global Reporting Initiative (2000–2010) Sustainability Reporting Guidelines; Mining and Metals Sector Supplement.

22 ICMM (2003) *Position Paper: Mining and Protected Areas* (www.icmm.com/document/43).

23 ICMM (2008) *Sustainable Development Framework: Assurance Procedure* (www.icmm.com/document/498).

24 International Finance Corporation (2011) *Performance Standard 6: Biodiversity Conservation and Sustainable Management of Living Natural Resources* (www.ifc.org/ifcext/sustainability.nsf/Content/EnvSocStandards).

25 IPIECA – The global oil and gas industry association for environmental and social issues (www.ipieca.org/focus-area/biodiversity).

26 LIFE Certification (2009) *Regulations for LIFE Certification: Preliminary Version.*

27 Millennium Ecosystem Assessment (2005) *Ecosystems and Human Well-being: Opportunities and Challenges for Business and Industry.* World Resources Institute, Washington, DC.

28 Fauna and Flora International (www.fauna-flora.org/initiatives/nvi/).

29 The Nature Conservancy (www.nature.org/aboutus/howwework/cbd/).

30 PricewaterhouseCoopers (www.pwc.co.uk/eng/issues/forest_finance_home.html).

31 Proteus (www.proteus.unep-wcmc.org).

32 The Wildlife Trusts' Biodiversity Benchmark (www.wildlifetrusts.org/index.php?section=corporate:biodiversity).

33 Hanson, C., Finisdore, J., Ranganthan, J. and Iceland, C. (2008) *The Corporate Ecosystem Services Review* (http://pdf.wri.org/corporate_ecosystem_services_review.pdf).

34 WBCSD (2011) *Guide to Corporate Ecosystem Valuation* (www.wbcsd.org/web/cev.htm).

35 CH2M HILL and WBCSD: *Global Water Tool* (www.wbcsd.org/web/watertool.htm).

36 PricewaterhouseCoopers (2011) for the World Economic Forum (www.pwc.co.uk/eng/publications/bio-positive.html).

Index